OUTDOOR
FLOORS

OUTDOOR FLOORS

The Design and Maintenance of Lawns, Patios and Ground Covers

PETER McHOY

Salem House Publishers
Topsfield, Massachusetts

First published in the United States by
Salem House Publishers, 1989
462 Boston Street
Topsfield, MA 01983

Designed and produced by
Breslich & Foss
Golden House
28-31 Great Pulteney Street
London W1R 3DD

Library of Congress Cataloging-in Publication Data

McHoy, Peter.
Outdoor floors: the design
and maintenance of lawns, patios and
ground covers/by Peter McHoy.
p. cm.
1. Lawns 2. Ground cover plants.
3. Patios – Design and construction. I. Title.
SB433.M34 1989
635.9′647–dc19 88-39907
ISBN 0-88162-408-X CIP

Printed and bound in Spain by Graficas Reunidas, Madrid

CONTENTS

DESIGNING THE GARDEN FLOOR

No matter how choice the trees and shrubs that you plant, how well planned your herbaceous and mixed borders, or how colourful the bedding plants, it's the floor of the garden that will make or mar the overall impression. Lawns, patio paving, even gravelled areas, act as a setting for the plants – and they emphasize the line and shape of the garden. However well you plant your beds and borders, they contribute surprisingly little to the sense of design, whereas the type of paving that you use, or the shape and quality of the lawn, immediately sets the tone of the garden.

It's probably the flower beds and borders that interest gardeners most, yet in terms of area they often represent only a minor part of the garden; they may occupy less than half the area of a small garden – probably a much smaller proportion in a large garden. For that reason alone, how you cover the remaining area of ground will have a major impact on both its appearance and the amount of time you have to spend on maintenance.

You can choose a hard surfacing material such as paving that will need practically no maintenance, or go for a high-class lawn that will look a treat but probably become the most time-consuming and labour-intensive element in the garden. But grass doesn't have to be mown every week in the summer, and paved areas needn't be boring or colourless; the clever combination of different surfacing textures will create interest, suitable plants around the edge can soften the hard lines, and planted containers will provide both height and colour.

Useful though the traditional grass and paving are, they're not the answer for all situations. There may be other plants, from fragrant herbs to prostrate conifers, that can be used instead of grass if you simply want a clothed area to look at rather than to walk on regularly. And there are hard surfaces that are just as practical, and sometimes more attractive, than clay or concrete paving – timber decking makes an interesting patio, gravel a super setting for plants.

The following three chapters deal in detail with all these alternatives. The first describes all the options for the lawn, from creating and maintaining grassed areas large or small, to 'lawn' plants other than grass; hard surfaces of all types are examined in the succeeding chapter; and finally suitable ground cover plants for beds, borders and banks. If you already know what you want, those chapters should give you inspiration along with the practical advice. If you've an open mind, however, this chapter may stimulate ideas, it should help you look objectively at the character and potential of the garden floor, and it will provide some solutions for problem areas.

Left: Grass will always remain the main surface for most gardeners – especially for a large area – and as a setting for beds and borders a lush lawn is difficult to beat. But for a large garden broad sweeps are generally more effective than formal rectangles.

Below: A 'wild' area (actually, you will need to sow the seeds of some wild plants to be sure of a good mixed display) is a low-maintenance solution for part of a large garden, with the bonus of the wildlife that goes with it. But bear in mind that it won't look so attractive in winter.
Opposite: Clay pavers are a smart way to provide an attractive sitting area and a natural transition from home to garden.

THE OVERALL DESIGN

If you're in the position of designing a new garden from scratch, and especially if money is no restraint, you're limited only by your imagination (and perhaps patience if you have to wait for the plants to grow). When an established garden has to be redesigned, there's always a temptation to tinker with the shape of the beds, do a bit of replanting, maybe add a raised bed and a few containers, and perhaps pave an area near the house and call it a patio.

It takes both courage and commitment to dig up an existing lawn, to lift a patch of paving and perhaps replace it with timber decking, or to turn part of a large lawn into a 'meadow' area, or to plant areas with non-grass ground cover. Yet if you simply let existing lawns or areas of paving dictate the profile of a new design, the chances are you won't radically improve it. Unless the features are really worth retaining, it's best to try to ignore them when you redesign the garden. It's often difficult, but sketching out your garden on paper, with only essential features drawn in to start with, is the way to come up with a radical and probably more effective design.

It's easy to dismiss 'design' as irrelevant if all you want is a simple, easy-to-maintain garden with the minimum of fuss; or you may assume that you need special artistic skills to be able to design your own garden. Neither is true, of course. Gardens can be designed to be labour-saving (many such gardens are illustrated in this book), and there's no reason why you shouldn't achieve a perfectly satisfactory design even if this is your first attempt – provided you follow a few basic rules.

Even a book on garden design can only give pointers and general principles, however. Good design is to an extent in the eye of the beholder, and a garden is a personal expression of what pleases the owner. But the following few tips may assist in taking an overall view of the garden, and help you to see the ground area as blocks or elements in an overall pattern before you become too deeply involved with the detail.

- At the early design stage, the choice of plants to use in a border is less important than its size, shape and proportion.
- Aim for simplicity: the plants will break up any sense of flatness and offset a rigid pattern.
- Design the garden first as blocks or shapes. Try not to think of these as 'lawn', 'paving', or 'flower bed' at

the early stage. Simply draw the basic outline on graph paper, and either sketch your patterns in pencil, or cut out pieces of coloured paper and move them around the plan.

- Use tracing paper over your drawn graph paper outline, to avoid having to keep redrawing it if you keep changing your mind!
- Remember that a balanced design does not have to be a regular one. You can have asymmetry and balance.
- Don't forget to try turning a rectangle at an angle to the house or fence.
- Try using circles, and intersecting circles.
- Make any sweeps and curves gentle, and always to a radius, even if different radii join in places.
- It is generally advisable to keep the design fairly formal near the house, and to make it more fluid and less formal further away.
- Avoid straight lines down a long, narrow garden; take some strong lines across the garden.
- Don't be afraid to divide a small garden up into even smaller segments, if this avoids the whole garden being taken in at a glance.
- Don't overlook the importance of a change of height. Even if the garden is on a flat site, decking or paving can often be raised slightly to provide more interest. A gravel area could lie a little below the level of the surrounding garden; a lawn could be raised a few inches above the surrounding area.
- When you arrive at a series of patterns, they must be translated into practical designs and modified to suit practical aspects like doors and gateways.

Opposite: Even a tiny lawn can be attractive and a real feature if it's distinctive enough. In this small cottage garden, a rectangular lawn of this size could have looked absurd, but this neat circular plot of grass becomes a focal point. **Right:** This garden was made for a disabled gardener, where ease of maintenance was important, but the principles are valid for anyone who wants a distinctive approach to gardening without having to spend much time actually at work.

Having established the basic shape of the garden, and determined the position of paths, consider the materials. Brick will give a very different 'feel' to, say, paving slabs; timber decking will probably make the garden look more 'designed' than concrete pavers. Above all, try to see these different textures and materials in association with each other, and with living carpets such as grass and other ground covers.

It is at the design stage that you can influence the amount of maintenance that will be required. Obviously, less lawn means less mowing, but there's the opportunity to use other attractive materials such as gravel or even old railway sleepers/ties, and beds can be carpeted mainly with ground cover, height being provided by a few choice trees or shrubs.

Sloping gardens may have steep banks that will suffer from soil erosion if not suitably planted, and they can be extremely difficult to work on (mowing may even be hazardous). Suitable ground cover can look attractive, save work, and avoid soil erosion. These are among the decisions best made at the design stage.

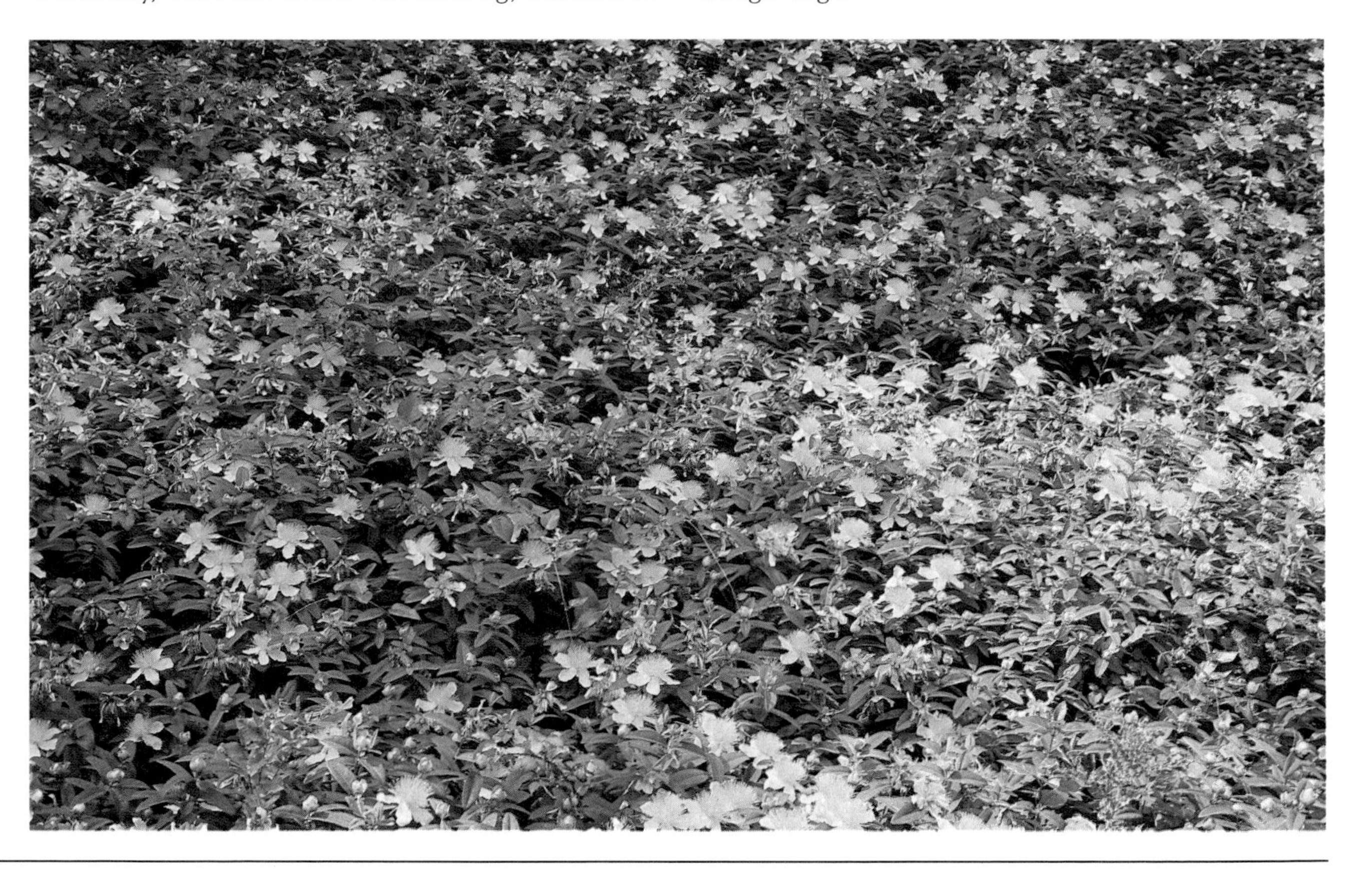

Hypericum calycinum, an ideal ground cover for slopes and other difficult areas.

Timber decking is an attractive option for a patio, especially where it can be linked with a water feature. Provided the wood has been suitably treated, it can be surprisingly durable.

IDEAS FOR SMALL GARDENS

It's traditional to have a lawn, but there are some superb gardens that have entirely dispensed with what can be a demanding feature. The smaller the garden, the easier it is to abandon the mower.

You may want to retain a lawn in the back garden simply because it provides a useful play area for children, quite apart from its generally decorative value, but if you have a small front garden, it's worth considering some of the alternatives.

A small rectangular lawn with borders around the edge, and possibly a bed in the middle, is not only unimaginative, it's also likely to be difficult to cut because mowing will involve a lot of short runs and a high proportion of edges in relation to lawn. So long as you think in terms of an area of grass, it will be difficult to give the garden much character.

Think of the whole area as a potential flower bed, then it may be possible to achieve something distinctive – and even labour-saving at the same time.

Below: It's worth making a special effort if you want a lot of impact from a small area. This Japanese style corner of a garden creates an intriguing effect that would be difficult to achieve with a traditional Western approach.
Opposite: A tiny front garden that would be completely unworthy of note if it had been grassed over or simply planted with a few shrubs. But the clever use of gravel and rocks has given it impact beyond its size.

If the area is bounded by, say, a drive on one side and a path on the other, a low-maintenance ground cover may give the same carpeting effect as grass, but with much more colour and interest. Bold drifts of heathers can provide colour for many months if you choose suitable varieties, and those with coloured foliage have a year-round appeal. You could use some of the sedums (but avoid *Sedum acre* and its variety 'Aureum', otherwise you'll rue the day you introduced it every time you have to try to weed it out from other parts of the garden). Prostrate conifers can also be very effective, and needn't be dull or uninteresting. There are plenty of other ideas for suitable plants starting on page 91.

Unless the area is very small, single-subject planting can be a bit monotonous, but using a few taller plants with ground cover between them gives you the best of both worlds. One of the most popular combinations is dwarf conifers (which have a tremendous variation in shape, height, and colour) with heaths and heathers carpeting between them. But there are plenty of other combinations to try, including the green-and-gold variegated *Euonymus fortunei* 'Emerald 'n' Gold' between red-stemmed dogwoods such as *Cornus alba* 'Sibirica'; in summer the leafy dogwood hides much of the euonymus, which nevertheless keeps down the weeds, but in winter the bright evergreen gold and green carpet makes a wonderful background against which to view the red stems of the dogwoods.

Gravel gardens are becoming more popular, and these can add a real touch of distinction. Gravel is cheap and easy to use as a form of hard surfacing and can be the solution to various problems of garden design. It's ideal where you want to fill an area with an irregular outline, or where paving would be too austere or harsh.

If the setting's suitable, a gravelled area can sometimes be given a Japanese effect with the careful use of votive lanterns, perhaps a few large but carefully chosen stones, and a few Japanese maples. This approach takes courage, but there are lots of other ways to use gravel (see page 77), most of which rely on it as a setting for plants. If you use small pea-sized shingle, alpines and dwarf shrubs may look best, but you may prefer an larger, more angular gravel, and then plants with a more dramatic profile may look better. Try yuccas, phormiums, and even angelica (which in its second year makes a large, striking plant).

You don't have to plant into gravel of course. It can simply take the place of the lawn. You'll need to use edging stones to hold the gravel in place, but a small lawn could be converted, retaining the flower beds and borders, with very little effort. It's worth considering if you find mowing too much of a chore.

Gravel can't compete with grass if you want a surface that you can lie on when the weather's hot, but it's almost maintenance-free and surprisingly attractive.

If you still want to make a feature of a lawn in a relatively small area, make sure it's well-maintained. It's worth spending money and time on it so that it really is an eyecatching feature. Nobody will give attention to a patch of second-rate grass, but closely cut, deep green, lush grass can be a focal point in its own right.

Even good turf will look better, however, if the lawn has a distinctive shape. A small circular lawn is a challenge to cut, but it won't go unnoticed or unadmired. This may not be practical in many gardens, but it's often feasible to turn a rectangular lawn at an angle of, say, 45°, so that it is not square to the house.

THE LARGER GARDEN

For a large garden, grass will almost inevitably be the major ground cover. It looks good, and with a powered mower it needn't be such a chore to cut.

Problems come only if you try to maintain the whole of a large lawn to a high standard. What looks bad is a large area of grass where none of it has been cut regularly – but you don't have to cut all of it regularly.

If you're bold and make the areas large enough, in generous sweeps, you can actually enhance what might otherwise be a rather monotonous expanse of grass by cutting it to different lengths. If you cut some parts perhaps twice a year, other areas perhaps once a month, and the rest say every week or ten days, during the cutting season, you will end up with a textured lawn that's surprisingly interesting as well as labour-saving.

The areas left to form 'meadow' will be rich with wild flowers, and your garden will be more interesting for the extra wildlife in general.

There may be areas that can be turned over to other ground cover plants, and this may be a far better solution for difficult positions, perhaps in shaded areas, where a good lawn is difficult to achieve. Ivies and pachysandras are among the plants that you can use to provide a low evergreen carpet over a large area that won't need cutting (many other suitable plants are suggested on pages 91 to 117).

Right: Large paths require an interesting surface if they are not to look boring. Crazy-paving always adds character, especially if natural stone can be used. **Opposite:** Large lawns look better if they are enclosed by shrub borders, and can be punctuated by focal points such as ornaments or, as in this case, a garden seat.

THE PATIO FLOOR

In Britain, relatively few gardeners cast their imaginations beyond concrete paving slabs. In many other European countries, and in the USA, gardeners are frequently more imaginative. Timber decking is more common, and even *in situ* concrete finds favour in some parts. This is not as unattractive as it sounds; tastefully done, with a textured finish, it can be surprisingly effective.

There is nothing wrong with concrete paving slabs, but they should be chosen carefully and used creatively (see page 64). Clay and concrete pavers offer plenty of scope, and they may be easier to lay as you can vibrate them into a bed of sand (see page 63). You can't do this with bricks (their dimensions are different), but they are versatile and can look very attractive. It's worth considering all these alternatives before making a choice.

Right: A change of level always makes a patio more interesting, and this is easy to achieve with timber decking.
Opposite: Never be afraid to mix surfaces – the effect is almost always better than can be obtained using just one material or finish, and it allows you to be more creative too.

Paving is often laid in a rather unimaginative way, and it may be possible to enhance it by using an attractive laying pattern (as in the examples of brick patterns on page 60). The laying patterns should be considered at the same time as the choice of material, as one can have an effect on the other.

No matter how suitable an individual paving material seems, it's likely to be even more effective used in combination with others. Try a row of bricks between every few rows of paving slabs, old railway sleepers/ties among bricks or pavers, cobbles with paving slabs, or maybe even areas of brick, natural stone slabs (or a good imitation), and railway sleepers/ties. All these combinations provide a contrast of texture and prevent the patio becoming commonplace or predictable. There are other materials that you can use, including fine gravel set in a glaze (see page 77), but it's not a good idea to use more than three different materials together, otherwise the patio can simply appear fussy.

There are probably some plant enthusiasts who regard hard surfaces, such as paving slabs, as anathema – man-made intrusions in a place set aside for things natural. While an almost completely paved garden, with only garden gnomes and the odd plant to provide relief, can only reinforce this view, hard landscaping can be an essential and integral part of good garden design.

Both paths and patios (or any area of paving) play a crucial role in giving a garden shape and form. They give it outline and structure, and are often the skeleton upon which the flesh of the garden is formed. A path in the wrong place, paving that is unsympathetic, will mar the garden, no matter how attractively you plant the beds or maintain the lawn. And the choice of paving will tell the world a lot about your taste and sense of design.

LAWNS

Lawns are frequently admired, but perhaps even more often cursed. They make a superb setting for colourful beds and borders, and if well-kept and a lush green are features in their own right. It's the demands they make on time and effort that sometimes make them unpopular.

If you really do find mowing a chore, even with modern equipment, it's best to look at alternatives to grass whenever possible. Once gardening becomes an unwelcome commitment, much of the pleasure is lost.

If the garden is really large, grass may be the only practical proposition for the parts where hard paving materials and ground cover plants would not be appropriate. The solution is to make the beds large, plant them with plenty of ground cover plants, use fairly coarse lawn grasses that still look good if cut by a large mower, and not to worry too much about weeds.

DECIDING WHAT YOU WANT

You can't take grass for granted. In northern Europe, grasses grow easily and they make a natural ground cover that can be depended on to look lush for most of the year. In these areas the problem is often one of restraining growth, which demands frequent mowing. In other parts of the world, perhaps where the climate is much warmer, colder, or drier, it is necessary to strive to get the grass to grow well enough to make a respectable lawn. A grass lawn always needs commitment. And the better the quality of the lawn, the more time and expense you will have to sacrifice. Settle for the type of lawn that you know you can maintain (or choose a non-grass alternative).

The quality of the lawn will dictate the amount of work that goes into preparation for sowing or turfing, as well as the amount of maintenance afterwards. An area required primarily for recreation (somewhere for the children to play, and perhaps a sitting out area), will be less demanding to maintain than a top-quality ornamental lawn intended to be part of the overall impact of the garden. Being realistic about expectations and commitments avoids disappointments.

If you want the kind of striped effect characteristic of the very best British lawns, expect to mow twice a week in summer, to feed the lawn regularly, use selective hormone weedkillers or lawn sand, maybe use moss controls, and to scarify, aerate and topdress at the appropriate times. Then to achieve the stripes you will also have to use a mower with a roller that will 'lay' the grass. And that presupposes that you started with suitable grasses.

Anyone dedicated enough will have a lawn the envy of their neighbours and it will be major focal point in the garden. But it will not be the kind of lawn

Left: A striped lawn is a traditional garden feature but a lawn like this requires a lot of dedication and work – and to achieve the striped effect you require a mower with a rear roller.

Below: On the small scale, a circular lawn is likely to look more interesting than a rectangular one. Here, the neat, crisp edge emphasizes the outline.
Opposite: This garden has many elements worth including in most designs: areas of hard (paving) and soft (grass) surfaces, a change of level, and strong focal points.

on which you want your children to play football. For a practical, hard-wearing lawn, stronger, coarser grasses, probably cut with a rotary mower without a roller, for speed and convenience, are a wiser choice. They will better stand up to the wear, and cutting once a week in summer is usually perfectly adequate.

A formal, perhaps rectangular, lawn looks better with fine grasses and a short cut; grass adjoining large flower beds and borders, and especially large lawns of informal shape, can be very pleasing with the less demanding, coarser grasses.

Weeds are almost certain to feature in a very large lawn (indeed, in times of drought they may help to keep it looking green). It's better to tolerate weeds than to worry about them.

These are mental adjustments that you have to make before you construct a lawn. It is only when achievements meet aspirations that the lawn becomes the object of satisfaction and not frustration.

ELEMENTS OF DESIGN

Formal lawns should reflect a formal design, but that does not mean rigid and unimaginative. Consider setting a rectangular lawn at an angle to the main axis of the house or garden – and don't overlook the possibility of a circular lawn. If the area is large, think about using a focal point such as a bird table, sundial, or ornament, to create a focal point.

Large, informal lawns should still have a purpose. At its simplest the lawn can lead to the end of the garden and the view beyond (great if you have a meadow at the end of your garden; not so good if you overlook a factory or the end of someone else's garden). But often it can meander to link one part of the garden with another – in effect like a very wide path. If the lawn does narrow where it links two separate parts or elements of the garden, it must be wide enough to spread the wear. If it's too narrow the constant walking over the same area will eventually spoil the lawn. Letting paving stones into the grass seldom looks right in this situation.

Many larger gardens have a big lawn surrounded by trees and shrub borders. These can often be made more interesting by leaving areas of grass long, like miniature meadows, where wild flowers can thrive. The long grass and wild flowers will also attract plenty of wildlife, and if the 'wild' areas are carefully curved into big islands and drifts, the whole lawn will have

Below: Buttercups and daisies – plants that most gardeners strive to eliminate from their lawns – can be surprisingly beautiful if you take a different approach to your gardening. 'Wild' corners reduce time spent on maintenance and can look good too.
Opposite: If a lawn is very large, leaving some areas long reduces mowing time and gives the area more shape and form.

much more character than a flat and rather boring expanse of evenly mown grass.

Large lawns also create plenty of scope for naturalizing bulbs, again possibly in areas left uncut until late in the year. The bulbs bring large drifts of interest in spring or early summer, and the foliage can die down naturally among the long grass and wild flowers of early summer.

Not everyone likes the idea of 'wild flower' or 'meadow' areas. If you're blunt enough to call them weeds, try the alternative approach of 'sculpturing' the grass by cutting different areas regularly but to different lengths. Use a very short cut for the main paths and walking areas, set in sweeps of intermediate-cut grass, and leave drifts of long-cut areas. By mowing the different parts say weekly, fortnightly, and perhaps every three weeks, with the mower set to different heights, interesting textured effects can be achieved that will again set your lawn apart from the ordinary.

If a relatively low-maintenance lawn is required, avoid complicated shapes that increase mowing time, and keep the number of flower beds within the lawn to a minimum (not only do these complicate mowing, they also increase the length of lawn edge to be trimmed).

IMPROVING WHAT YOU HAVE

Most established gardens already have a lawn. The chances are it is a rectangle, possibly with a few beds cut into it. And unless it was originally sown with fine grasses, and has been well maintained since, it won't really be possible to make it into a first-class lawn. It is, however, an existing asset that you can almost certainly improve.

Digging up an existing lawn is a major decision. It's hard work too. Sometimes it is necessary to make such tough decisions to get the garden looking right. Just as a straight path down the garden will dominate the design no matter how you try to vary the flower beds, so a lawn will set the overall tone of the garden and affect the design fundamentally. Designing around existing paths and lawns seldom works: if a strong sense of design is important to you, it's best to plan your garden on paper without regard to existing lawns. But if all you want is a green and pleasant patch on which to relax, then modifying what you've got has a lot to commend it.

Changing the shape can often be effective. Would it look more imposing as a circular lawn? Could you make it into an L-shape simply by lifting the turf from one corner and relaying it along one of the other

edges? Would a paved area around the edge give it a crisper and more designed look? Could you reshape it so that it makes a rectangle that lies diagonally across the garden instead of parallel to the house? This may be possible by lifting some of the turf and relaying it. All these simple steps can improve the design element easily and quickly, and the results are instant.

Changes of shape are best achieved by cutting into the existing lawn, and if necessary relaying the turf to add to another area. Sowing an additional area of grass, perhaps to fill in an existing flower bed cut into the lawn, is seldom completely successful. Unless the grass seed mixture used is very similar to the original mix, the new area will look different in shade or texture. If you do want to try extending the lawn with new seed or turf, at least match the type of grass mixture – use a ryegrass mixture if it is a fairly coarse lawn, a fine lawn mixture (without ryegrass) for a lawn composed of the choicer grasses. But don't expect an exact mix (even different varieties of the same species can vary in colour). Of course, if you happen to live in a part of the world where single species lawns are used, you must use the same grass.

RENOVATING AN OLD LAWN

Even quite neglected lawns can be improved with time and effort. Just routine lawncare attention, applied more radically, may be all that the grass needs to bring it up to scratch.

If you have inherited a lawn that has regularly been cut too short or too long, simply adjusting the mowing height for a month or two will help to bring colour and vigour back into the grass. Set the blades to cut at about 1in (25mm), and don't cut closer until the grass is looking healthy again.

Apply a lawn fertilizer appropriate to the season (one high in nitrogen in spring and summer, low in nitrogen and high in phosphates in autumn). And if the weather is dry, water thoroughly to get the grass growing vigorously again.

Weeds are best tackled once the grass is looking good, and is able to compete and colonize the spaces left by the weeds that have been killed.

It's worth checking the soil pH and if necessary adjusting it to discourage weeds and encourage the finer grasses (see pages 30 and 83). But make sure you take samples from the lawn itself and not the surrounding beds.

With a neglected lawn, the chances are a 'thatch' of dead grass has built up at soil level, and the ground itself has become compacted. Scarifying and aerating (see page 46), perhaps followed by a top dressing, will usually improve things dramatically over a period of several months.

A really mossy lawn is a big problem. The steps suggested above will all help, and the use of moss killers can certainly achieve a short-term improvement. But the real problem may be poor drainage or too much shade, and these have to be tackled if a long-term solution is sought. It may be better to start again, with improved drainage and a more open site.

Bumps and hollows can often be levelled very successfully provided they are not too extreme (see page 52).

Once the grass seems to be growing healthily, and preferably in spring or early summer, get rid of as many weeds as possible. Broad-leaved weeds, such as daisies, are easy to eliminate with selective hormone weedkillers. Coarse grasses, such as Yorkshire fog (*Holcus lanatus*), or Dallis grass if you live in the States, will remain to mar the lawn. If there are just a few clumps of these difficult and unwanted perennial grasses, try painting them carefully with glyphosate. *This will kill all the grasses that it comes into contact with,*

A NEW LAWN: TURF OR SEED?

The case for turf	• Instant results • Easier for someone without experience to get good results • Can be laid at any time that the ground is not frozen
The case against	• Generally more expensive than seed • Unless you pay a lot for it, or are careful to see what you are going to get before you buy, there is a significant risk of getting inferior turf • Less choice of specific grass seed mixtures; this probably won't matter if you want a general-purpose lawn, but could if you have, say, a shade problem, or want grasses that will perhaps withstand hot, dry conditions
The case for seed	• A wide choice of grass seed mixtures of known suitability • Generally cheaper than turf
The case against	• More difficult for an inexperienced gardener to get good results (ground needs very careful preparation and needs careful watering and nurturing until established) • Should only be sown at certain times of the year (usually spring or autumn, when the weather is warm enough to germinate the seeds, but the ground not so dry and hot that the seedlings suffer)

so use it carefully. Even if you try to paint the gel formulation carefully onto the leaves, you will almost certainly kill some of the surrounding desirable grasses, but the patch can be reseeded.

Once the perennial weeds have been killed:

- Cut the grass low.
- Use a powered lawnrake to scarify the surface and remove the thatch of dead grass.
- Go over the lawn with an aerator, rake up and remove soil-cores from a hollow-tined aerator, and then brush in a mixture of equal parts peat and sand (or fine soil).
- Loosen the soil with a hand fork or trowel where there are large bare patches.
- Sow a suitable seed mixture at twice the rate recommended for a new lawn (many fewer seeds will germinate and grow because many will not be in contact with the soil).
- Use a sprinkler to wash the seeds into contact with the ground, and use as often as necessary to keep the soil moist until the new grass has germinated. This is essential to success and may be necessary daily (or twice a day) in very hot or dry weather.
- Once the new grasses are growing, cut the lawn with the mower set at about 1in (25mm), and use a lawn fertilizer to stimulate growth.

A NEW LAWN

There used to be a clear-cut reason for using either turf or seed for a new lawn: turf was instant but you usually had to make do with inferior grasses; seed was slow, but it was the only way of being sure of a top-quality lawn containing only the finest grasses.

Nowadays you can buy turf grown from specially sown mixtures that are as good as any that you can buy and sow yourself, but you may find it more expensive than a carpet for your home.

If you can afford to buy specially grown turf (that is turf specially sown for the purpose, and cut very thinly when 'harvested'), this will give you the very best of both worlds: good turf and instant results. This kind of turf can be bought in varying grades, and some of the mixtures are perfectly suitable for a general hard-wearing lawn.

If you can't afford this kind of turf, the choice lies between 'ordinary' turf (which may be cut from specially prepared meadows, in which most of the weeds have already been eliminated), and seed. There are a few other minor options, of minority appeal, mentioned later.

PREPARING THE GROUND

Laying turf does not let you off preparing the ground. It's just that preparation has to be that bit more thorough if you are sowing seeds. Unless the ground has been consolidated and levelled properly, hills and dales will be as obvious if you lay turf as if you sow seed.

The main objectives are levelling and eradication of deep-rooted perennial weeds that will become a problem later. But improving the structure of the soil (by giving it better moisture-holding properties or perhaps by improving the drainage) will make maintenance easier and improve the quality of the grass.

If the ground is very weedy (especially if it has not been cultivated for some years), it's sensible to use a weedkiller. It will make the digging easier if the top growth is cleared first. Don't use persistent weedkillers for this job – glyphosate is ideal because it will be translocated within the plants to kill the roots of perennials, yet you can replant right away (though for this to work properly you should leave the weeds for a week or two for the top growth to die and the roots to be killed).

If the ground is not too weedy, it's best to do the initial digging and levelling first, which will expose more weed seeds. Leaving the ground for a week or two will allow these to germinate, and give time for missed roots of perennials to start growing. You can easily hoe off all the seedlings, and the few perennial roots can be dug up. If you prefer spraying to hoeing, you can use a general weedkiller for killing top growth, leaving the ground available for immediate replanting.

Most lawn weeds can be controlled at a later stage by modern selective weedkillers, but there is no point in starting off at a disadvantage. And although selective weedkillers can be used once the grass is established, the weeds will be competing for light, moisture, and nutrients during those crucial early weeks.

Anybody who has had to lay land drains will confirm that it's a tiresome and tiring job, to be avoided if possible. If the drainage problem is slight, it may be adequate to lay the whole lawn on a slight slope. But if there is a high water table, or if there is a real drainage problem with water usually lying on the surface after rain, one simply has to face up to the task ahead. Trying to install a drainage system after the lawn has been laid is a major task, and the grass probably won't look the same again.

Provided you have used a weedkiller if the ground is infested with difficult perennial weeds, the easiest way to prepare a large area is with a rotary cultivator (well worth hiring – you are unlikely to need it for more than a day). For a small lawn, digging is a

LAYING DRAINS

Traditional land drains are made of fired clay, but there are plastic equivalents that are lighter to handle. Unfortunately the hard work lies in excavating the trenches. The best time to assess whether poor drainage is a problem is during the winter months, but drainage improvements are best carried out from spring to autumn, when the soil is drier and easier to work.

Lay the main straight drain with a gentle fall leading to a runoff at the lowest point. Subsidiary drains (above right) should radiate from the main drain herringbone-fashion. As drains can soon become clogged, cover joints with a small square of plastic sheeting, and backfill with gravel (below right) before finishing with soil.

There are now modern plastic-based land drains with a narrow profile, about 4in (10cm) deep, but only ¾in (2cm) thick. These smaller types require less excavation and are the best choice if you have to lay drains in an existing lawn.

Follow the manufacturer's advice regarding spacing, but for lawns the drains may have to be about 3ft (1m) apart on heavy soils, perhaps only 20ft (6m) apart on light soil.

To install these narrow profile drainage systems in an existing lawn, remove a strip of turf at least 4in (10cm) wide, though the width of a spade may be more convenient. Remove the turf evenly, and then excavate a trench about 9in (23cm) deep (lay a sheet of plastic on the grass alongside the trench, and put the soil on this). Lay the drain along one side of the trench, return the soil, firm it with a heel, then return the turf. Water thoroughly if the weather is dry.

Main drain and subsidiary

The covered joint

practical proposition. On good soil, it will probably be adequate to fork over the ground, but if you garden on heavy clay, shallow chalk soil, or a light soil with a hard pan (a layer of consolidated ground at the depth of normal cultivation), or if drainage is poor, double digging (see below) will be worth the effort.

If major levelling is necessary, be sure to keep the less fertile subsoil at the lower level, even if this means removing some of the topsoil, levelling the subsoil, and then returning the topsoil. This is easy to recommend, a major and exhausting task if you have to do it by hand. It may be worth employing a contractor with suitable equipment to do the job. An alternative, if the site is suitable, is to import some topsoil and spread this over the area to achieve a level (but make sure you buy good topsoil).

Generally, however, the necessary levelling is on a much smaller scale – ground that looks almost level to the eye, but with small hills and valleys that become obvious once you start checking with a level (see opposite).

Whether sowing seed or laying turf, the soil should be reasonably consolidated before levelling. This does not mean using a large, heavy roller. Just tread the ground firmly, shuffling your feet along, before raking level.

If the soil is in poor physical condition (too sandy and dry, or heavy clay, for example), at least try to improve the top few inches by incorporating plenty of peat or sand, or anything that will improve the structure without making the job of levelling difficult (clumps of manure for instance).

If you want a really good lawn, it is worth having a soil analysis made (you can test the pH yourself, but a nutrient analysis is generally best done by a soil testing laboratory). If you don't want to bother with this, spread a balanced general fertilizer into the top layer of soil at the rate of about 4oz to a sq yd (140g to a sq m): but do not add fertilizer if you are preparing the lawn in late autumn or winter.

All this should be done a couple of weeks before making the lawn, to allow time for weed seedlings to germinate, and the soils to settle.

DOUBLE DIGGING

Double digging is only necessary where drainage has to be improved, or where you want to improve the structure of difficult soils. It is unnecessary for most lawns.

Start by taking out a trench about 2ft (60cm) wide and 10in (25cm) deep.

Fork over the bottom of the trench, loosening any compacted soil. Some gardeners dig manure or garden compost into the bottom of the trench, but for lawns and shallow-rooting plants it is best worked into the top 10in (25cm) instead.

Excavate the next trench, throwing the inverted clods of soil forward to fill the previous trench.

Repeat the process until the end is reached, then barrow the soil excavated from the first trench to fill in the last.

TESTING THE pH

The degree of acidity or alkalinity of the soil is expressed on a pH scale that ranges from 1-14, with 7 as neutral. This can be a bit misleading because the scale is logarithmic and and a pH of 8 is ten times more alkaline than one of 7, so a few points on the scale can represent a big difference.

It is always worth checking the pH of your soil, and

LEVELLING THE GROUND

Make a set of pegs by cutting a piece of timber into lengths of about 8-10in (20-25cm), with the tops sawn absolutely straight. Mark around each peg with a pencil 1in (2.5cm) from the top, using a square. Taper the other end for easy insertion into the soil.

Drive the pegs into the roughly levelled ground about 3ft (1m) apart. Peg one section at a time and as one part is levelled, move the pegs to extend the area.

Use a spirit-level on a straight-edge to align the pegs to the same height (below left). You may need to hammer some in, or re-insert others more shallowly. The pencil marks indicate the final soil level.

If the ground is not too uneven, you should be able to rake it level (below right). Large discrepancies may require the use of a spade.

there are simple methods that you can use at home. The meters are convenient and easy to use, but not always accurate unless carefully chosen and used with care. The colour comparison kits are reliable enough for general purposes.

The exact method varies slightly from one kit to another, but the principles are the same.

Take a sample of soil from the top 2-3in (5-8cm), dry it a little if it's very wet, and remove any stones or fibrous material. Mix this with the indicator fluid, and once the particles have settled compare the colour of the liquid with the colour chart provided. With some kits the liquid is drawn through a filter; some of the more expensive kits have chemicals that speed the settling process.

A LAWN FROM SEED

Sowing the seed is not the most important part of growing a new lawn. What comes before and after is just as vital: the choice of suitable seed and the care of the seedbed while the seeds are germinating and becoming established.

Unless you are an expert in grasses you will have to depend on the descriptions of the mixtures provided by seedsmen, and your best guarantee of a good mixture is to buy from a very reputable source.

You can usually tell whether a mixture is intended primarily as a quality ornamental lawn or as a hard-wearing utility lawn by whether it contains ryegrass. Mixtures with this grass are likely to be hard-wearing and tough, if lacking the quality of the finer grasses. Those without ryegrass are likely to be intended for admiring more than using.

This generalization is less true than it used to be. Some of the modern varieties of ryegrass are compact, with horizontal rather than upright growth and much finer leaves than the old varieties. It comes as a

The effect of different frequencies of feeding and weeding on the same grasses can be seen in these trial plots. Both colour and texture can be affected.

surprise to many gardeners that there are in fact many different varieties of some of the common species of grass, but much breeding work goes into producing strains with better growth habit, better colour, and attributes like disease and drought resistance. Even if the names of the varieties used in the mixtures are given, they are likely to mean little to the gardener, and new ones are coming along all the time. That's why it is worth placing your trust in the name and reputation of the supplier.

Most seed suppliers offer mixtures for shade and nearly all can provide mixtures containing one or more varieties of perennial ryegrass together with, say, a little creeping red fescue and a Chewing's fescue variety for tough play areas; and there will probably be a fine turf mixture of just Chewing's fescue and brown top *(Agrostis tenuis)*. There will be others containing different percentages and perhaps some different grasses.

Although mixtures are almost always sown, in the southern States lawns may be prepared from single species, such as Bermuda grass *(Cynodon dactylon)*. Climate plays an important role in the choice of grasses – in the USA and Canada there are seven lawn grass zones, each with its own general growing conditions and grasses adapted to them. In Britain the country can be taken as a whole, and the mixtures sold should do well anywhere in the area.

Sowing

It's not a good idea to use rule-of-thumb application rates for grass seed – follow the seedsman's recommended rates. The number of seeds to a gram varies greatly from species to species: perennial ryegrass may have as many as 500 seeds in one gram, some of the bents *(Agrostis)* over 15,000 so the number of seedlings you are likely to get from a given weight depends on the seeds in the mixture. The different seed counts should also be borne in mind when looking at percentages in a mixture: a seemingly low percentage of one grass may actually represent many more seeds than another with a higher percentage by weight.

It should not be necessary to 'add a little extra for the birds' – the recommended rate will allow for some losses. If birds are a particular problem in your area, you can sometimes buy seed treated to make it unpalatable to birds. If you don't realize until after you have sown the seed, black cotton stretched across the area, and the use of bird scarers, should minimize the problem. Bird scarers should be moved about regularly if they are to be effective.

Even application is as important as the rate. The best method is to use a mechanical spreader. Some, which you can also use for spreading fertilizers, are pushed along the ground, others (cheaper and just as effective for this job) are hand-held and the seed scattered by cranking a handle. With both of them, it's worth having a trial run on a path, or on sheets of newspaper spread over the ground, so that you can gather up the seed afterwards and weigh it to check that the rate of distribution is reasonably accurate.

A few seedsmen provide the seeds in 'scatter boxes'. You simply tear off a strip along the edge of the packet, which exposes holes through which you can scatter the seeds as you walk along.

If you don't have a spreader, just follow the step-by-step guide opposite. It's a bit more bother, but will work just as well.

AFTERCARE

The priority is always to keep the ground moist. Seeds won't germinate if the ground is dry, and the seedlings can easily be killed if subjected to a dry spell soon after they have germinated. It may be necessary to use the sprinkler a couple of times a day if the lawn is sown during the summer.

If you don't have a sprinkler that will water the lawn from a position at the side (often impossible if the lawn is large), put the sprayer and hose in position and leave if there, just connecting it to the tap as necessary. In this way you avoid walking on the ground too often, and grass will soon grow to fill in the bare area left by the hose once it has been removed.

Wait until the grass is about 3in (8cm) high before attempting to cut it, and then keep the mower blades sharp. Do not take more than 1in (2.5cm) off for the first few cuts.

It's a good idea to give the lawn a light roll with a garden roller (or with a rear-roller mower, with the blades set very high), before you start mowing properly. This will help to firm the plantlets into the soil so that they are less likely to be damaged by the first cut, and it will press any loose stones into the surface and so reduce the chance of the blades being damaged. As the rolling will flatten the grass, wait for a day to give it the chance to return to the upright position again before cutting.

If you have a grass catcher, use it while the new lawn is becoming established, even if you don't use it later. It will reduce the chance of the seedling grasses becoming smothered.

Apply a lawn fertilizer once the grass has been growing for a month or two (but not until late spring if you sow in the autumn). Don't use a selective weedkiller until the grass has been growing for at least

SOWING GRASS SEED

1 Make sure the ground is level (see page 31), and use a rake to break up large clods of earth and remove stones and debris.

2 Tread the soil firmly, shuffling the feet, then rake it again to produce a smooth, level surface. Unless the ground has settled significantly it should not be necessary to level with pegs again.

3 Use a garden line to mark yard or metre strips. If you are not confident about sowing evenly, use half the recommended quantity at first, then apply the other half in strips the opposite way.

4 Finally, rake the seed in lightly, removing foot-marks as you go. Water thoroughly with a sprinkler (not a hose-pipe, which will wash the seeds into drifts or down cracks in the soil).

LAYING TURF

1 Start by laying two straight edges. If you can make the lawn in multiples of the turf size, it is best to work from these two edges.

2 Using a plank to stand and kneel on if possible, to distribute your weight, lay the turf row by row, staggering the joints like brickwork.

3 Firm the sods to make sure they are in good contact with the soil underneath. A 'turf beater' is sometimes used for this, and you can make one yourself out of a slightly wedge-shaped piece of wood.

4 Fill in any gaps between the joints with fine, dry soil, or a mixture of peat and sand. Go over the area with a tamping tool (the back of a spade if nothing better), if not already 'beaten' as laid.

six months. If you notice any difficult-to-eradicate weeds becoming established (such as clover), either pull them out by hand, or paint the individual plants with glyphosate, or use a hormone weeding stick that you simply wipe over the leaves.

A LAWN FROM TURF

Although a lawn from turf is instant in its effect, it takes more physical effort and time to lay than it does to sow seed. As compensation, the ground preparation does not have to be quite so thorough – although a level surface and freedom from deep-rooted perennial weeds are still a prerequisite.

It can also be laid successfully at any time of the year, provided the ground is not frozen. If you can water regularly, and take precautions to prevent the turf from drying out before you lay it, summer is an ideal time as the grass grows away quickly. If there is any doubt about the ability to water regularly until established, early autumn to early spring is a better time.

Unless you know the reputation of the supplier is good, and you have seen the quality of his turf before, always make a point of inspecting the grass before you buy if this is at all possible. Because of high transport costs, you will generally have to buy from a fairly local supplier anyway. You can buy turf from some garden centres, but rolled or stacked turf has a limited life once cut, and it pays to be cautious about buying this way unless you know the turf has only very recently been delivered.

Always try to lay within a couple of days of delivery. This means having your ground ready before you order (or specify delivery after you know that the ground will be ready). If the weather prevents immediate action, unroll each piece in a shady position, and keep watered.

Difficult Shapes

If you have to fit an area where the last edge would comprise narrow strips (which are difficult to keep in place, and dry out quickly), lay all the edges first, and work inwards from the edges, cutting the sods to fit in the centre.

Don't attempt to lay turf to an intricate shape, or incorporate curves at this stage. Lay to rectangles first, then cut the lawn to shape afterwards.

Aftercare

Unless it rains steadily after laying, put the sprinkler on. A light shower will probably only wet the surface and not penetrate to the crucial layer beneath where you want the grass to root into the moist soil below.

Although turf is not as sensitive to drying out as a newly-sown lawn, be prepared to water thoroughly whenever the weather is dry, at least for the first few months.

Newly laid turf should be mown after about four to six weeks. But keep the grass fairly long, about 1in (2.5cm) at first. If you buy good quality turf, weeds will not be a problem. But if you do need to apply a selective weedkiller it should be safe to do so once the lawn has been down for a month or two and the grass is growing actively (but also follow the manufacturer's instructions regarding best times for application).

OTHER METHODS

Although in Britain, lawns are almost always created from seed or turf, in warmer climates and in parts of the USA, sprigs and plugs are used. Sprigs did have a spate of popularity in Britain, but the lawns were not ideal for hard wear (partly because they were composed of just the one species – usually a creeping bent), and heavy raking or rough mowing tended to drag up some of the mat of grass.

Sprigs, which are very successful for grasses such as Bermuda grass *(Cynodon dactylon)* in the southern part of the USA, and which may sometimes be offered in Europe, are quite easy to plant.

The sprigs may be sold in airtight bags to keep them moist; unwrapped sprigs should be kept damp until planted. Prepare the ground as for seed, but water thoroughly a day before planting. Use a draw hoe to create shallow furrows about 3in (8cm) deep and 6-12in (15-30cm) apart (closer planting means quicker cover). Plant the sprigs 6in (15cm) apart, holding them against one side of the furrow while drawing soil over the roots and firming (make sure some of the leaves are above soil level). Water thoroughly, and hand weed or hoe as necessary until the plants clothe the area, which may take six months or a year.

Plugs (small pieces of turf) are easier to handle than sprigs. You can use a trowel to plant them, but where available a plugging tool that takes out a soil core of the right size is much more efficient. Take out the holes 6-12in (15-30cm) apart, fill each one with water, let it soak away, and then plant the plugs, firming in so that the top of the plug is level with the surrounding soil. In the warm areas where plugs are sold, they should provide ground cover after about a year. In the meantime, hand weed or hoe as necessary.

Pre-seeded rolls make an appearance from time to time. These have the seeds embedded in a biodegradable base rather like thick paper. They are simply cut to size, put in position, covered with a thin layer of sieved soil, and watered. Very easy to use, but unless you take the precaution of pegging the strips down they can be ripped up by a high wind. Another method that has been used is to start the seeds off on rolls of hessian, grown by hydroculture. This type of 'lawn on a roll' (whether seeds or growing grass) is particularly useful for steep slopes, where it is difficult

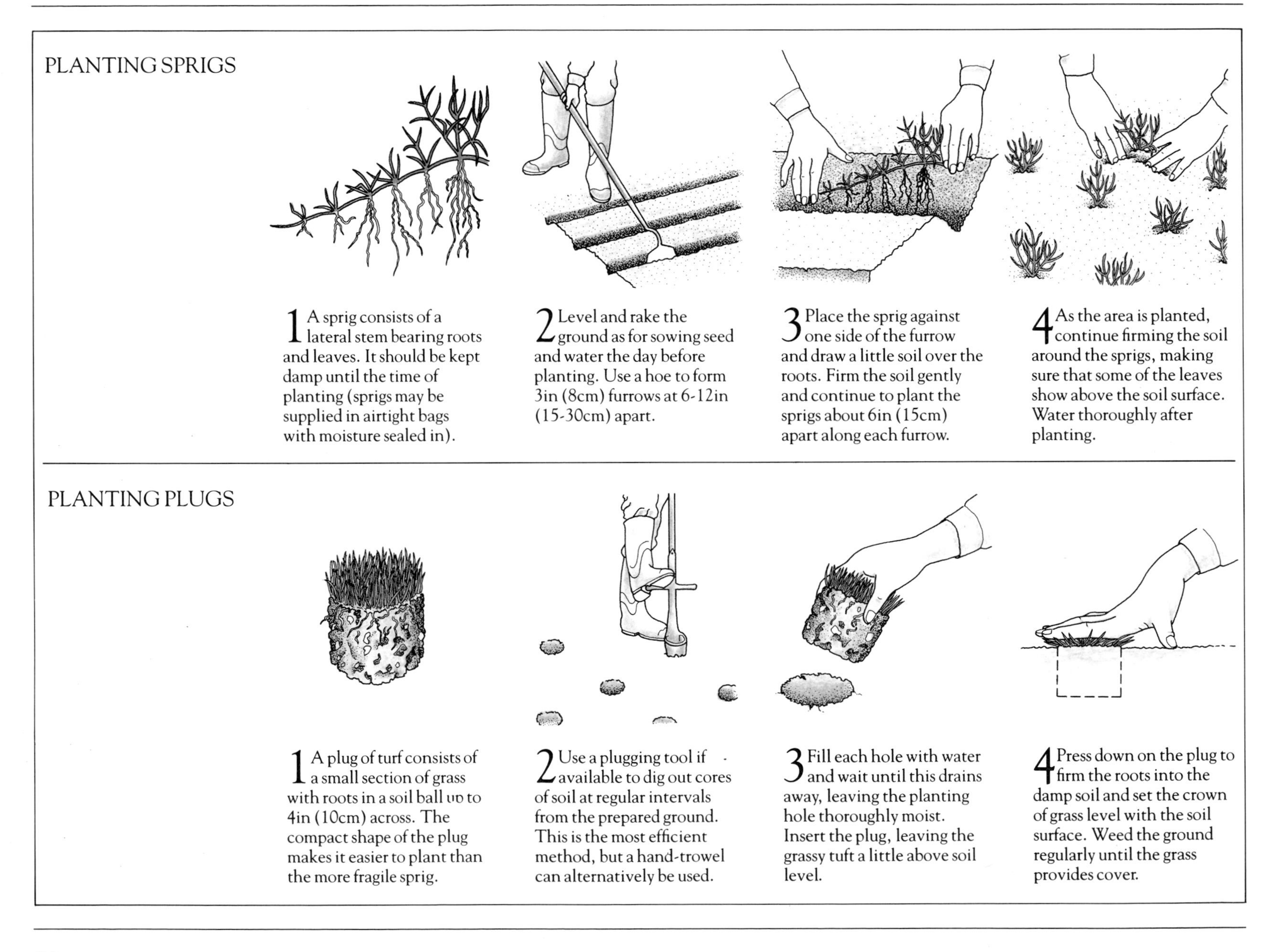

PLANTING SPRIGS

1 A sprig consists of a lateral stem bearing roots and leaves. It should be kept damp until the time of planting (sprigs may be supplied in airtight bags with moisture sealed in).

2 Level and rake the ground as for sowing seed and water the day before planting. Use a hoe to form 3in (8cm) furrows at 6-12in (15-30cm) apart.

3 Place the sprig against one side of the furrow and draw a little soil over the roots. Firm the soil gently and continue to plant the sprigs about 6in (15cm) apart along each furrow.

4 As the area is planted, continue firming the soil around the sprigs, making sure that some of the leaves show above the soil surface. Water thoroughly after planting.

PLANTING PLUGS

1 A plug of turf consists of a small section of grass with roots in a soil ball up to 4in (10cm) across. The compact shape of the plug makes it easier to plant than the more fragile sprig.

2 Use a plugging tool if available to dig out cores of soil at regular intervals from the prepared ground. This is the most efficient method, but a hand-trowel can alternatively be used.

3 Fill each hole with water and wait until this drains away, leaving the planting hole thoroughly moist. Insert the plug, leaving the grassy tuft a little above soil level.

4 Press down on the plug to firm the roots into the damp soil and set the crown of grass level with the soil surface. Weed the ground regularly until the grass provides cover.

Although pre-seeded rolls are not particularly popular, they are sometimes available and do make sowing quite easy. They are of practical value for sloping sites, where seed may otherwise be washed away after heavy rain.

to sow seed easily (even if you sow evenly, the seeds are often washed into drifts by the rain), but is not a serious contender for most lawns.

ALTERNATIVES TO GRASS

Plants intended primarily as practical and decorative ground cover but not usually regarded as a lawn substitute are dealt with in the chapter on ground covers (page 79) but there are a few plants that are sometimes grown in place of grass.

Chamomile and thyme are two popular lawn substitutes, and moss is sometimes used. There are lesser-known lawn substitutes, useful for the right situation, but none of them as versatile and hard-wearing as grass for most sites.

The following list describes plants suitable as a direct substitute for grass, together with a practical assessment of their merits and limitations.

Chamomile *(Anthemis nobilis)* has been used as a grass substitute for centuries, albeit on a very limited scale and successfully probably in just a few famous gardens. It sounds appealing: it's fragrant and there's no need to mow. The negative side is that you can't use selective weedkillers as you can with grass, so weeding has to be done by hand, it can be walked on but won't stand up to heavy wear like grass, and it can be difficult to establish. There's also the risk of patches dying out.

It can't really be recommended for situations where you would normally have a utility lawn, but it's worth a try for awkward areas that would be difficult to mow, such as perhaps a narrow border along the edge of a path.

Chamomile has small feathery, aromatic leaves and white daisy-type flowers, and spreads rapidly by creeping stems. If you choose the ordinary form that flowers, it will require trimming, but the non-flowering 'Treneague' should not. The ordinary species can be raised easily from seed, but you will have to buy off-shoots of 'Treneague'. These will probably arrive in a plastic bag, and you will have to plant them about 6in (15cm) apart. Planting them closer may give cover a little more quickly, but it will work out more expensive if you have to buy off-shoots, and the final cover may be no better. Although you can plant in autumn or spring, spring is the best season, to allow plants time to establish and

Although grass is often the first living carpet that one thinks of, there are many more unusual options. Here are just four of them.
Far left: Carpet of moss in a Japanese-style garden.
Top left: Chamomile (*Anthemis nobilis*), one of the most popular grass substitutes.
Centre left: A carpet of creeping thyme (*Thymus serpyllum*) and in the background *Acaena glabra*.
Below left: *Cotula squalida*, a fairly rampant plant that can produce an effective carpet.

reduce the chance of winter losses.

If growing chamomile from seed, sow under glass in early spring, and plant out in late spring.

Provided the plants are well watered until established, they should cover the ground after about three months, during which time you will have to resort to hand weeding, though if difficult grasses such as couch (*Agropyron repens*) are a problem, a selective grass weedkiller like alloxydim-sodium can be used.

Chamomile needs coaxing to grow well. Try topdressing with sieved soil or a mixture of equal parts sand and peat, in mid autumn, and in mid spring apply a balanced general fertilizer. Trimming with the mower set high will encourage the development of sideshoots if the plants do not seem to be bushing out well enough. Flowering strains will need mowing occasionally to prevent flowering.

Don't bother with chamomile if the soil is poorly drained, and wet in winter. And if the soil is very alkaline, work in a thick layer of peat before planting. If the soil is not to the plant's liking, you may have to replant after two or three years.

Creeping thyme is another aromatic plant that will tolerate a little trampling, but don't expect too much from it. Be sure to use the creeping thyme (*Thymus serpyllum*) and not the bushier *T. vulgaris* (the one usually used as a herb).

Thymes are good on dry soils, and tolerate alkaline conditions well, but become straggly after four or five years and may then have to be replanted.

There are other thymes that could be used, but varieties with more decorative foliage are probably best used in other ways.

Cotula is sometimes used in place of grass in New Zealand, but in Britain C. *potentilliana* and C. *squalida* are sometimes regarded as weeds in fine turf. Try them, provided you are prepared for them to be rampant spreaders that may need curtailing. They are low-growing plants with fern-like leaves and small yellow flowers.

Mosses can be very attractive. They make a good substitute for grass beneath trees. In the USA, some of the plants used under the name of moss are not true mosses, but species of *Arenaria* and *Sagina*. These are known in America as Irish and Scotch mosses – pretty hummock or mat-forming plants, but not such practical carpeters as some of the true mosses.

The problem with using moss is that you can't buy it from a nursery in the way that you do other plants. You have to provide the right conditions and then encourage native mosses, transplanting them if necessary. You need a moist, acid soil, and if necessary you may have to apply sulphur to achieve this (see page 83). Transplant moss clumps in early spring and be sure to water for the next few weeks.

The main trouble with moss, which otherwise does well in the shade beneath trees, is the falling leaves – these have to be raked up (without pulling up the moss with them), as the moss will die if buried.

If you have a really mossy lawn and don't mind experimenting, you could try converting it to a moss 'lawn'. By making the soil acid (down to a pH of 5.5 or less), you will encourage the moss and discourage the grass. If other conditions are right the moss will eventually take over, and you will have a lawn that is green for most of the year, that needs no mowing or fertilizing, and which will soon bounce back again even if it goes brown during a period of drought.

Other plants are sometimes suggested, including marjoram (*Origanum vulgare*), pennyroyal (*Mentha pulegium*), and even pearlwort (*Sagina procumbens*). But these are not without their drawbacks and if you are looking for a practical substitute for grass rather than a novelty, they are best avoided.

A LAWNCARE CALENDAR

SPRING

Start mowing in earnest. You should have been mowing occasionally in mild spells during the winter if the grass was long, but spring is the time when you must act to prevent the winter growth becoming too long. Start as soon as the risk of heavy snowfall and harsh frosts has passed.

Feed if the grass looks yellowish. If you regularly remove your clippings, the grass will almost certainly require a spring feed to replace lost nutrients. But wait until mid or late spring, when the weather is obviously improving and the grass growing vigorously again, before feeding. Lush growth stimulated too early may be damaged by late harsh frosts.

Using a fertilizer spreader will make the job easier, otherwise mark the lawn off with strips 1yd (1m) wide as you work along the lawn, and divide this visually into 1yd (1m) sections as you apply the fertilizer. Use some kind of measure into which you can weigh the correct amount for 1sq yd (1sq m) before you start, then use this as a guide each time.

Always water in lawn fertilizers unless it is raining, otherwise they may scorch the grass.

Scarify and aerate if this was not done in the autumn, and the grass needs it. These are traditionally autumn jobs, but they can be done just as well in spring. High quality lawns are often scarified more than once in a season anyway.

Topdress if there are hollows to be levelled, or after aeration. Again this is traditionally an autumn job, but spring is also a suitable time.

Moss control is often effective if carried out in spring (see page 50). Lawn sand is a method that you can use in spring.

Weedkilling with selective weedkillers is generally best done in late spring or early summer, when the weeds are beginning to grow vigorously. Don't cut the grass for a few days before you apply a selective weedkiller, as you want a good area of leaf for the chemical to act upon. Wait to cut the grass for about four days after applying the weedkiller.

To roll or not to roll? Gardeners previously rolled their lawns quite frequently, but nowadays few gardeners possess a roller and the lawns are no worse for that. If the ground has been lifted by winter frosts, rolling in spring will help to knit the lawn together again, but in milder areas and on heavy clay soils, rolling is likely to harm through compaction rather than help. If you do roll the lawn, a garden roller of 3-5cwt (150-250kg) is adequate.

SUMMER

Feed the grass if you haven't already done it in the spring. Even if you have, the grass may benefit from an additional high-nitrogen feed during the summer. If you feed in spring, the second application should be made in mid summer.

You can overdo the fertilizers. Too much growth can simply mean more frequent mowing to keep the lawn looking good. On the other hand a high-nitrogen fertilizer will make the grass look greener.

You will be able to find dozens of brands of lawn fertilizer. Some may do better on one type of soil than on another. Without access to comparative trials, it is difficult to work out which is the best one to buy. If you find one that you are pleased with, it makes sense to keep to that one.

There are simple alternatives that will be cheaper, and will probably give you satisfactory results on most lawns. Straight sulphate of ammonia (ammonium sulphate) can be used at a rate of ½oz per sq yd (15g per sq m), every couple of months during the summer.

Mix it with about four times its weight of sharp sand to ease spreading, adjusting the application rate accordingly. This is a good option if you are trying to make the ground more acid. Lawn sand (see moss killing, page 50), can also be used.

If you want to reduce costs by using just one fertilizer for most of the garden (cheaper because you can buy in bulk), don't be afraid to use a general balanced plant food (it might be lower in nitrogen than lawn feeds, but perfectly satisfactory for balanced healthy growth). In Britain, for example, Growmore (which is a formula and not a brand) is one of the best and most economical general fertilizers that you can buy.

The formulation may also influence your buying decision. Most lawn fertilizers are sold as fine powders (which are sometimes described as 'microgranules'), to be applied dry. These are less likely to scorch the grass if you don't water them in thoroughly enough, but can be unpleasant to apply on a windy day. Coarse granules are easy to apply, but may scorch the grass if not watered in very thoroughly. Some come as liquids, which you dilute with water. A few can be used in hose attachments, which makes application easy but you have to be careful to apply the right amount. Those that have to be applied by watering-can can be tiring and tedious to apply unless the area is small.

Watering is the most important summer job. But to be really effective you must water *before* the grass has started to turn brown. Once the grass looks dull and seems to have stopped growing, and the turf feels less springy underfoot, it is time to start watering. For the very best lawns you may have to water regularly from late spring to autumn, especially on light, sandy soils.

Frequent light watering is almost useless. It just encourages surface rooting, rendering the plants even more vulnerable to drought, so the effect is more obvious as soon as you have to stop watering for any reason (perhaps a holiday). For thorough watering, you will need to use a sprinkler (see pages 47-48).

Aim to apply about 1in (25mm) or 5 gallons to a sq yd (25 litres to a sq m) each week if there is no appreciable rain. This is best done in two applications of half this amount. You can place a rain gauge in the lawn to check how much you are applying. But as some sprinklers do not apply water evenly, it may be easier to place a number of straight-sided containers around the lawn, and water until the average depth of water is about ½in (12mm). You only need to do this measuring once if you time how long it takes to achieve this coverage – next time just leave the sprinkler on for the same length of time.

It's best to water early in the morning or in the evening, then less will be lost through evaporation.

Mowing is a job that everyone knows is a major preoccupation throughout the summer months. Here are a few tips to keep it in perspective:

- For an ordinary domestic lawn, mow once a week, with the cutting height set to about 1in (25mm). If you have a high quality lawn composed only of fine grasses, reduce this to about ¾in (18mm). Very high quality lawns are cut shorter – to about 5⁄16in (8mm); professionals go even closer during the summer, but cutting as close as this must be done frequently (several times a week), with plenty of feeding and other routine care. Close mowing also encourages moss.

If you leave the grass to become too long, it will be difficult to cut, the results will look patchy, and the grass may look discoloured after cutting.

- If you have a large lawn, try cutting areas of it less frequently, with the blades set higher. This will create

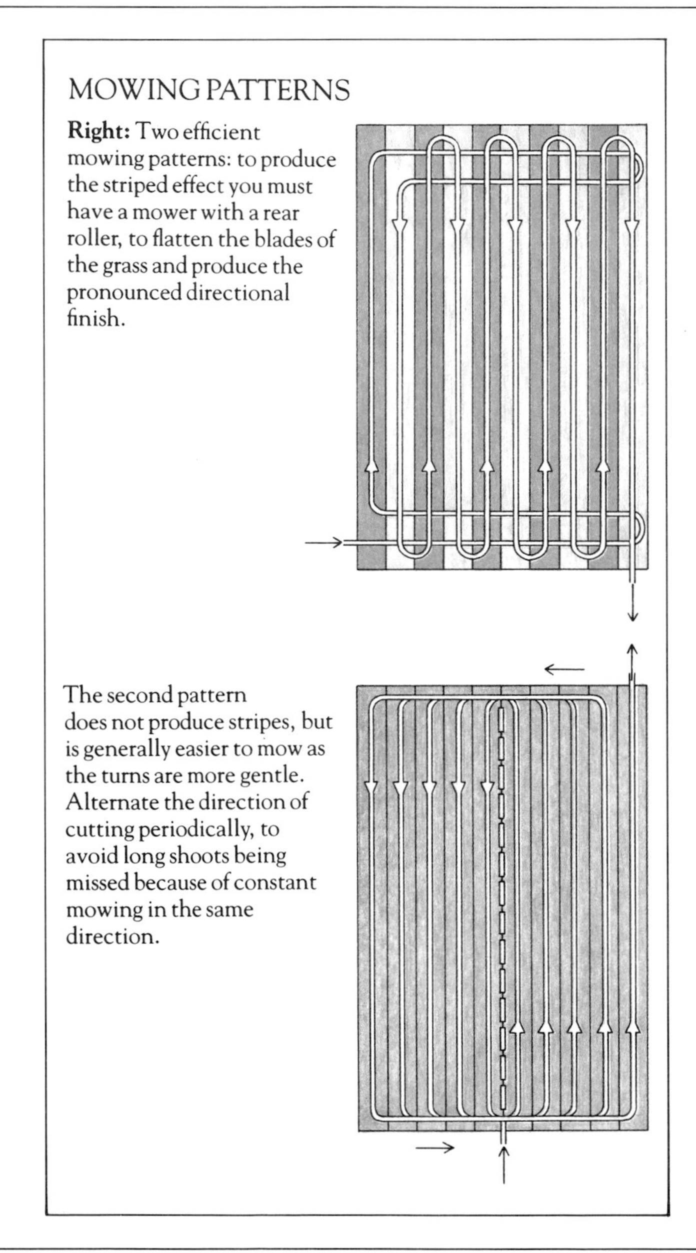

MOWING PATTERNS

Right: Two efficient mowing patterns: to produce the striped effect you must have a mower with a rear roller, to flatten the blades of the grass and produce the pronounced directional finish.

The second pattern does not produce stripes, but is generally easier to mow as the turns are more gentle. Alternate the direction of cutting periodically, to avoid long shoots being missed because of constant mowing in the same direction.

a textured effect with areas cut to different heights (effective on a large lawn, but not on a small one).

- Don't bother to collect the clippings (except for the first cut of the season, or if the grass has been left to get long). Short clippings will not contribute significantly to a thatch, and the nutrients will be returned to the soil.

AUTUMN

Continue to mow as long as the grass is still growing. Once very cold and frosty weather arrives, you can clean the mower and put it away for a while (though this is a good time to get it serviced).

If you neglect mowing, coarse grasses will be encouraged at the expense of the finer ones. This is especially important in high quality lawns.

Raking can be hard work, but it does pay dividends. If you have a cylinder mower with a front roller, the longer, creeping stems may be flattened and not cut – raking lifts them and generally helps to prevent a tangled, matted surface.

A thatch can also build up, sometimes quite thick, from fibre and stems at the base of the plants, along with decaying grass clippings. This tends to encourage waterlogging after heavy rain, can affect root growth and the vigour of the plants, and may encourage moss.

Raking gets rid of the thatch, may help to control moss, and will also discourage trailing weeds such as clover and chickweed. It is possible that it may spread the moss if you don't improve conditions overall.

If you have a large lawn, consider a powered lawn rake. They really are effective, and make light work of a tiring job. If you don't consider it worth buying one, it may be possible to hire one.

Aeration is useful if the lawn is suffering from excessive compaction, or if surface drainage is poor.

It's most likely to be necessary on heavy soils or lawns that take a lot of wear. High quality lawns are spiked several times a year, but once a year, in autumn, is a good compromise for most lawns. Lawns on light soils, or that take little wear, may not require aerating more often than every two or three years (though as it's a tedious job you could treat just a portion of the grass every year).

Clear leaves off the lawn as quickly as possible. If left to accumulate they will prevent light reaching the grass, leading to yellow, even dead, patches. They can also encourage fungus diseases to spread if left to lie on the grass.

This is a once-a-week job if you have trees around your lawn, possibly more often if big leaves such as sycamore are a particular problem. A stiff brush is adequate for relatively few leaves or a small lawn, and some hand lawn rakes are very effective. Hand-pushed or powered sweepers are worth considering if you have a very large area to cover (these sweep the leaves up into a collecting box or bag).

Topdressing can be done in spring, but autumn is also a good time. If you are planning to aerate the lawn, do that first.

Topdressing is perhaps the most misunderstood aspect of lawncare – it's done regularly by professionals, seldom by amateurs, to whom it often remains something of a mystery.

The vast majority of lawns are never topdressed, and for everyday use will probably be perfectly adequate. Topdressing is something to consider for that special lawn, or to remedy a particular problem.

There's no one formula for topdressing. You can use sieved garden soil (if it contains a few weed seeds, it will probably be cheaper to kill these if they grow than to buy sterilized soil). Peat is sometimes used, and can be helpful on dry, sandy soils, or in alkaline gardens (but avoid coarse peat with large pieces, as this can be difficult to work into the surface).

Sand is often the choice for heavy soils, such as clay, or a mixture of equal parts sand and sieved soil. Don't use builder's sand (it is generally too fine, and may also be alkaline).

Topdressing is useful on all lawns where you want to raise the level of a hollow, but always do it in stages – don't try to raise the level by more than about ½in (12mm) within a season. Usually it is used to improve the soil structure and to stimulate root growth, in which case it should be used after aeration, so that the topdressing can be brushed down into the holes or slits.

Scatter the material with a spade first, broadcasting it as well as possible, then use a stiff brush or a lawn rake to work it well down among the grass.

Feeding should not be necessary if you fed the grass in spring and summer. But if the grass looks weak and poor by early autumn, try using an autumn lawn feed. Autumn lawn fertilizers differ from spring and summer feeds by having much less nitrogen (which would stimulate soft, sappy growth liable to be damaged in the winter). Lots of moss and yellowish grass are signs that an autumn feed may be worth while.

Moss killers can be used successfully in the autumn (see page 50).

WINTER

Cut the grass in mild spells if it seems to be getting long, but set the blades high. If you let the grass become too long, when you eventually cut it in spring the finer species may have suffered, and bare patches may develop.

Keep off the grass when it is wet or the weather very cold.

LAWNCARE TOOLS

The lawnmower Surely the greatest and most-used mechanical aid invented for the gardener, it's the one almost every gardener owns, and certainly the one it would be the most difficult to do without. For that reason lawnmowers represent big business, and they are heavily advertised by major manufacturers.

The choice can be bewildering, and the merits claimed for one type or another by manufacturers do not always help. Now, with some rotary mowers having rollers, and others having grass collection facilities, the once clear-cut relative merits are becoming blurred.

Width of cut has an important influence on how long it takes to mow the lawn. Most cut about 1ft (30cm) at one pass, which is adequate for small lawns. A wider cut saves time, but possibly at increased effort (especially in the case of hand mowers).

For a medium-sized lawn up to 600sq yd (500sq m), it is worth choosing a mower with a cutting width of 14-18in (35-45cm). For very large lawns, a cutting width of 24in (60cm) may be more appropriate.

Final choice of model must depend on personal preferences, requirements, and how much you can afford. But it is possible to narrow the field to one or two types by asking yourself a series of questions. Use the guide on the opposite page to help you choose the type most likely to be suitable.

Hand rear-roller mowers are good for a striped finish on a small lawn, where you don't mind the effort involved in pushing a fairly heavy machine. They don't cut long stalks as well as rotary mowers.

Hand side-wheel mowers are easier to push than the rear-roller type, but you won't get a striped finish, and it isn't easy to cut up to the edge of a flower bed (the side wheel will dip over the edge).

Power cylinder mowers have the merit of a striped

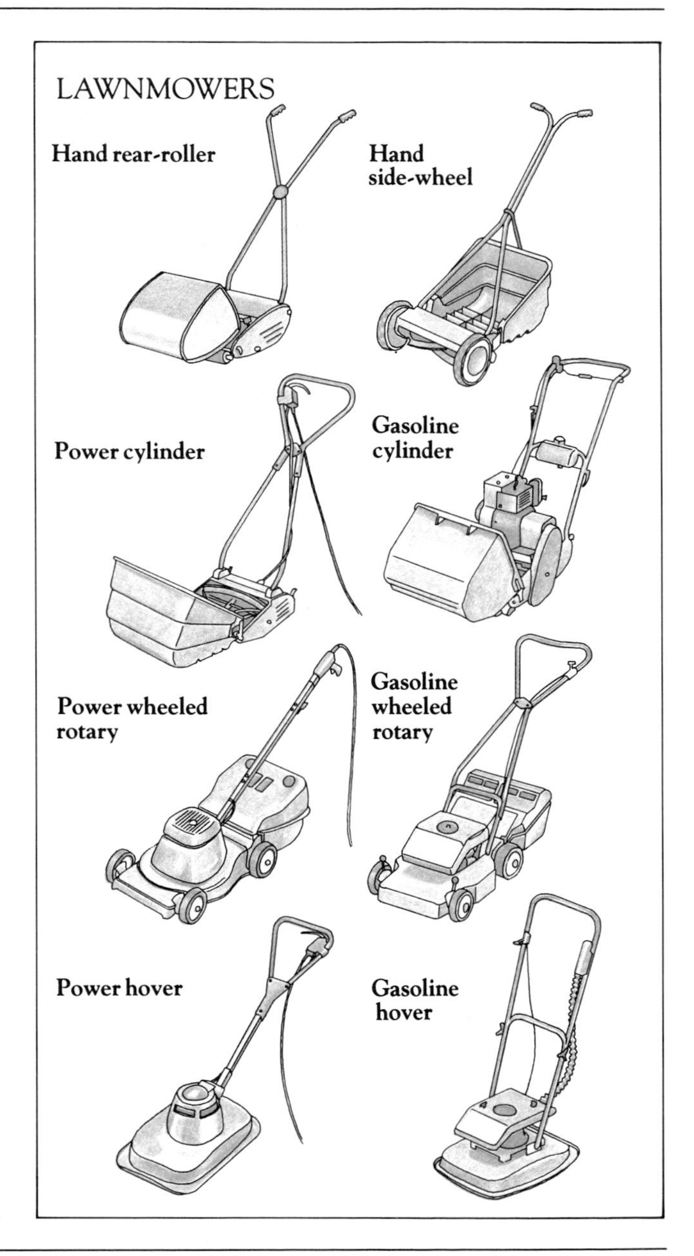

CHOOSING A MOWER

If you're in doubt about the most suitable type of mower for your lawn, this simple guide should help you. Just answer 'yes' or 'no' to the questions below, and the key will lead you to the mower that's most likely to be suitable for your requirements.

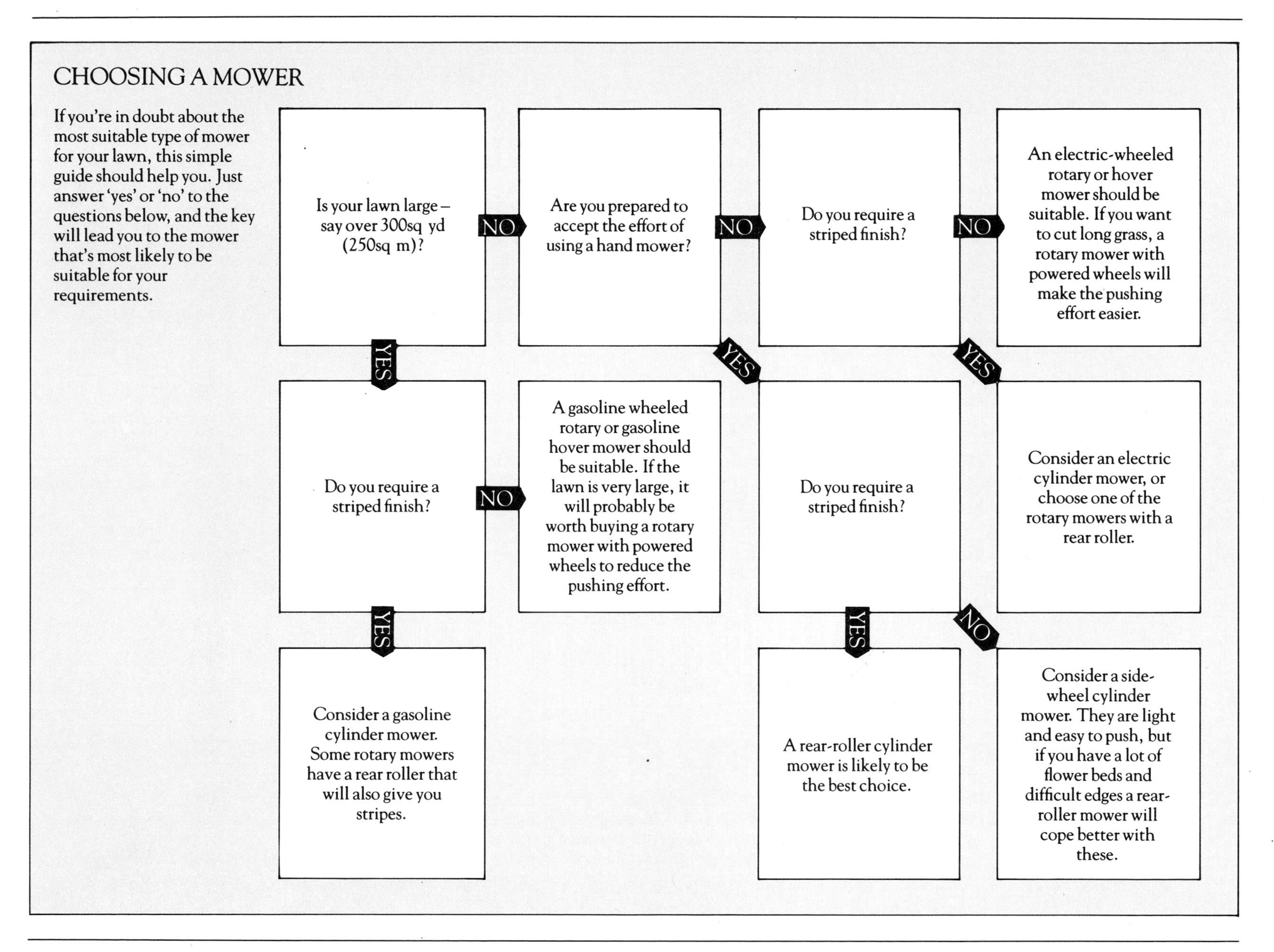

finish with the minimum of effort. But again they will not deal with long stalks as well as a rotary mower, and in comparison with gasoline or hand mowers you have the drawback of a trailing cable.

Gasoline cylinder mowers are ideal for a large lawn (where a trailing cable and a power supply could be a problem), and if you want a striped finish. But they are noisy, and much more expensive.

Power wheeled rotary mowers are fine for small lawns if you don't want stripes. There are few disadvantages for a small lawn, where a power supply and trailing cable are relatively minor drawbacks.

Gasoline wheeled rotary mowers are suitable for a large garden, and where a striped finish is not important (though some do now have rear rollers to achieve this), and they generally deal well with both long and short grass.

Power hover mowers have the merit of being very easy to handle, and they make mowing a fairly effortless job. But you won't have a striped finish, and they can be dusty to use in dry weather. You will also have to be careful at the edges of flower beds, which will be scalped, and the mower blade possibly damaged, if you go over and lose the cushion of air.

Gasoline hover mowers have most of the pros and cons of the power type, but you have freedom from the restraints of a power supply – although a model of this type comes with an increased price tag and more noise.

Lawn rakes are useful for clearing up leaves and raking up clippings as well as for scarifying the surface and loosening the thatch. Most traditional lawn rakes have metal 'teeth' with a degree of springiness (they are called spring-tined rakes), arranged in a fan shape. You can also buy plastic versions (not so suitable for scarifying of course).

A useful tool, but if you have a large lawn, a powered lawn rake is a much better proposition.

Flat-tined rakes, which can otherwise resemble spring-tined rakes, are intended primarily for clearing leaves and clippings.

Slitter rakes are intended to slit or scarify the ground more than spring-tined rakes, but they are sometimes hard and tiring to use. Again, a powered lawn rake is worth thinking about.

Aerating tools These items probably won't be worth buying unless you have a large lawn and are likely to aerate at least a couple of times a year. It might be best to hire one for a day.

Some have spikes or slitters on a revolving drum or wheel (you can also buy attachments to fix to the front of some mowers). These can be hard to push, and although they will help to improve drainage, do not do such an efficient job as a hollow-tined aerator for overcoming the problem of compaction.

Hollow-tined aerators take out cores of soil (which should be swept up if there is not a collecting box). They are driven in to the ground (usually by pushing on a handle while pressing with a foot), to remove a core of soil. However, unless the soil is moist, it will be very hard work and you probably won't succeed in removing whole cores.

Slitters will cut through the thatch but won't penetrate enough to do much for compaction.

Solid tines are easier to push into dry or hard ground. A garden fork can be used instead of a special aerator, but it is hard work and it's best to do a small area each week or month, rather than to attempt the whole lawn at once.

Fertilizer spreaders Although probably not worthwhile for a small lawn, a fertilizer spreader is useful for

LAWNCARE TOOLS

Three simple but very useful and effective lawncare tools. Lawn rakes (right) are inexpensive and invaluable for removing moss and loosening the 'thatch' that otherwise builds up and causes the lawn to deteriorate. Hollow-tined aerators (below) can be tiring to use, but do help to improve the turf. A fertilizer spreader (below right) will speed up the job, and by making it easier may encourage you to feed the grass more often.

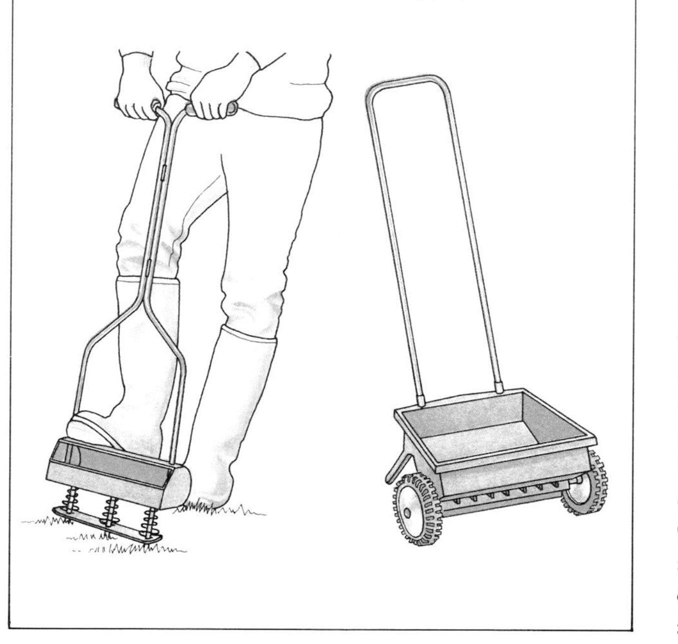

a large area – and of course you can probably use it for sowing grass seed, and weeding too if you use a granular combined weed and feed. How well it works depends a lot on the model that you buy. Not all will cope with all types of fertilizer equally well, and some may not be much use for dressings such as lawn sand.

Wheeled spreaders should be wide enough not to make the job tedious: a 16in (40cm) spread should be adequate for most lawns. It should not be too heavy (bear in mind the weight when filled), and it's useful if it has an on/off control so that the contents don't spill out when you don't want them to – when you are wheeling it from one part of the garden to another, for instance.

A choice of settings may enable some to deliver half the dose in one direction, then the other half in the other direction – a way to ensure even coverage.

There are a few hand-held shakers and distributors (you either crank a handle or shake the container). These have the drawback of being tiring to use over a large area, unpredictable in delivery rate (it depends on your walking pace, and the vigour of your cranking or shaking); and you'll probably get a fair amount of the contents over your legs and shoes.

Sprinklers There are four main types of sprinkler:

Static sprinklers are the simplest and cheapest. You simply push them into the ground and they produce a circular pattern of water over a fixed area (a few will cover rectangles). The area covered is generally small, and you will have to move the sprinkler about to cover even a moderately sized lawn. Most water is delivered close to the centre of the circle.

Oscillating sprinklers are more versatile. They throw a fan of water over a rectangle, and the size and area can usually be adjusted by dial (water pressure also affects the area covered). Coverage is likely to be fairly even, though more will be delivered close to the sprayer.

Rotating sprinklers have nozzles that rotate, producing a circular pattern. They are likely to cover a larger area than a static sprinkler, but cost more.

Pulse jet sprinklers cover a large circular area. They are expensive, but the best choice for a large lawn. The jet can be adjusted to cover areas of varying size.

The illustrations below show each of these sprinklers and their watering patterns.

TROUBLESHOOTING

It is beyond the scope of this book to deal with individual pest and diseases, or even to give a comprehensive list of lawn weeds. To make identification meaningful would require a whole book on the subject, and different pests and diseases are problems in different places. While in Britain the problem is leatherjackets (the larvae of the daddy-long-legs or cranefly), in the States it is commonly sod

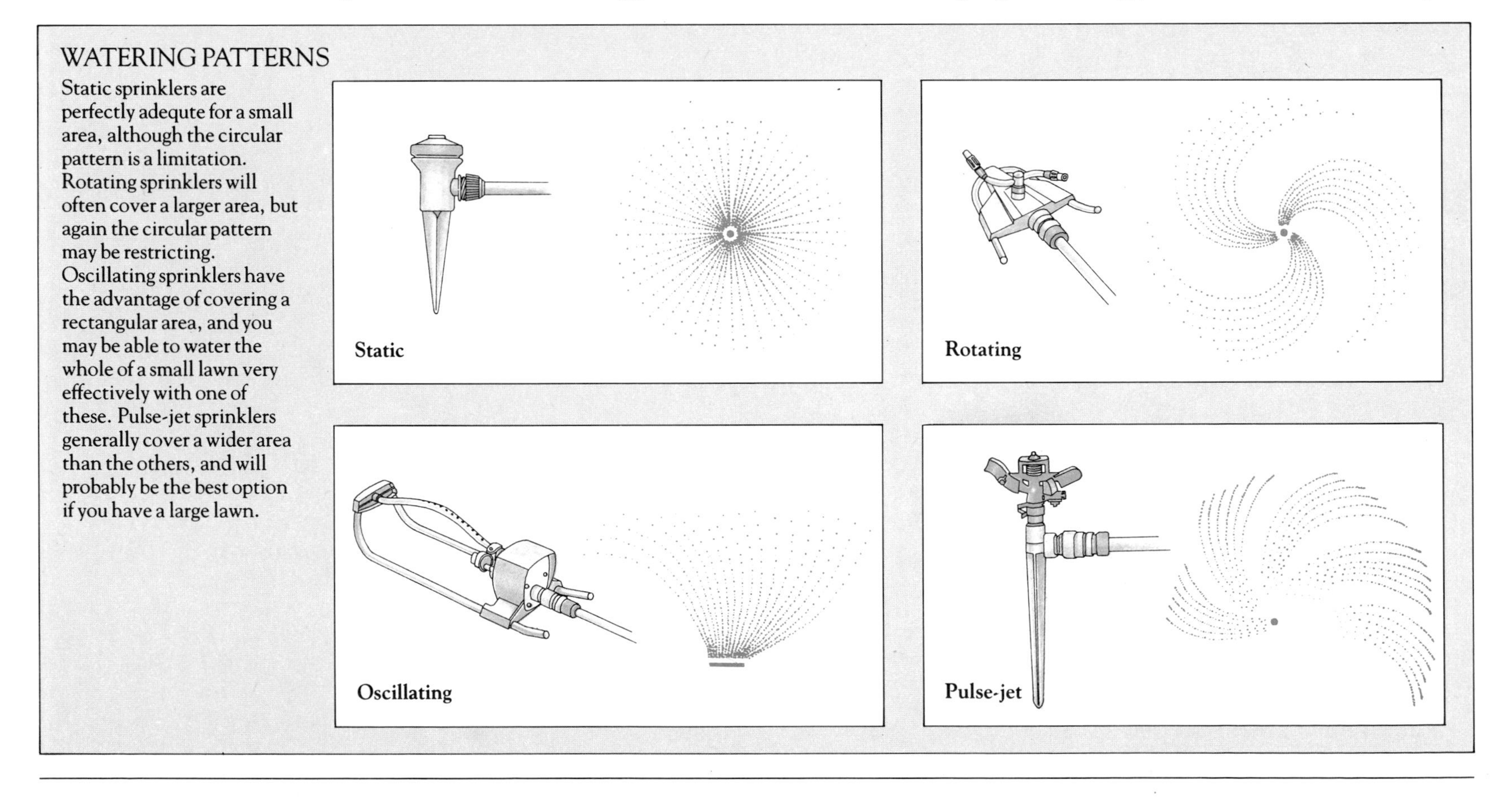

WATERING PATTERNS

Static sprinklers are perfectly adequte for a small area, although the circular pattern is a limitation. Rotating sprinklers will often cover a larger area, but again the circular pattern may be restricting. Oscillating sprinklers have the advantage of covering a rectangular area, and you may be able to water the whole of a small lawn very effectively with one of these. Pulse-jet sprinklers generally cover a wider area than the others, and will probably be the best option if you have a large lawn.

webworms (the larvae of a small moth) that cause concern. New pesticides (and new brand names) are sure to appear between the writing of this book and its publication (and certainly during the time that it is on your bookshelves). And of course chemicals that are available in one country may be banned or simply not available in another. New research may lead unexpectedly to the withdrawal of chemicals that we all thought were part of the lawncare scene: ionynil is an example. This previously widely used selective weedkiller was suddenly withdrawn in Britain in 1985.

That does not mean you can't do some meaningful troubleshooting, and take steps that will reduce or eliminate problems. Many can be solved by cultural means, with others you may need to resort to widely available chemical controls. You don't always need to identify the problem exactly to know what to do. If grubs are chewing the roots and killing the grass, it doesn't matter a lot which grub it is: control is likely to be similar for most of them. Most fungus disease of lawns can be controlled with a fungicide, such as benomyl or thiophanate-methyl: by no means would this be the preferred treatment of all of them, but the instructions on the product will give an indication of which diseases are likely to be controlled.

Right: A clear demonstration of the power of selective (hormone) weedkillers. The strip on the right shows the result after ony 10 days.

Weeds

If you can live with a few weeds in your lawn, it will save a lot of frustration. But clearly if you are aiming for a really first-rate show lawn you don't want *any* weeds.

Weeds are less likely to be a problem if the soil is acid rather than alkaline, and if the grass is well fed and looked after. Mowing can help too: cutting a fine lawn to about ¾in (18mm) and one with coarser grasses to about 1in (25mm) will help to control weeds if you start with clean turf.

Hand weeding is immediate with results, and the least expensive method. If you want to use selective weedkillers, you don't have to treat the whole lawn if only part is affected. Treat just that area with a weedkiller that you water on, or if there are just a few plants, use an aerosol or stick that you wipe on the leaves of the weeds. These are useful for keeping on top of any new weeds that happen to appear.

Not all selective weedkillers kill all the different kinds of weeds equally well, though with repeated applications you should win through. Many proprietary brands contain cocktails of several hormone weedkillers, to kill as many different weeds as possible. Brands and formulations change, so check the label to see which weeds a given formula is claimed to control (though obviously if you have some obscure weeds, don't expect to find them mentioned on the label). You may not manage to control the most difficult weeds within a single season, but if just a few are left try treating them with a spot weeder.

Selective weedkillers sold for use on lawns will not kill weed grasses. That general statement may be challenged by some gardeners who use selective weedkillers to kill certain grass species on certain types of established grass lawns, but unless you are absolutely sure that the product will kill only the grasses that you want destroyed among your particular lawn grasses, it's best not to try.

A few very coarse grasses that form a clump of tough, spreading leaves can be dug out, or weakened by repeated slashing. If you are very careful you may be able to paint glyphosate onto the leaves without touching the other grasses. You will be left with a small patch that may require reseeding. Coarse grasses with an upright habit can usually be eliminated eventually by close mowing.

Moss is often much more difficult to eradicate than other kinds of lawn weeds. Even if you kill off the plants, the spores will germinate unless you use a product that will kill those too. And unless you improve the underlying conditions that encourage moss, it will probably return.

Start by improving drainage, eliminating as much shade as possible (perhaps by cutting back overhanging shrubs or removing a branch from a tree if it can be done without spoiling the shape of the tree), and by feeding the grass. Get the mowing height right too: if you mow too close to the ground, cushion-forming mosses will be encouraged; if you leave the grass too long, the feathery trailing mosses will flourish.

Moss killers will help in the short term.

Lawn sand applied in early autumn, with a follow-

AN ALTERNATIVE TO CHEMICALS

Some gardeners dislike the use – or over-use – of chemicals in the garden, and if you go by the book you could be applying a lot of chemicals of different kinds to your grass.

It doesn't have to be that way. You can have a respectable lawn without using an armoury of chemical weapons.

Here are the basic rules:

- Don't cut the grass too short – try to avoid removing more than a third of the height at each cut. Little and often is better than taking a lot off at once.
- Prevent a thatch building up by scarifying the surface at least once a year (once a month is not too often during the growing season).
- Aerate if the grass looks in need of it (see page 46).
- Topdress annually.
- Leave the clippings on the lawn, to recycle the nutrients.
- Dig out isolated weeds by hand.
- If there are lots of weeds, consider living with them (daisies can be attractive), or use a weedkiller to get the lawn clear initially, then fall back on hand weeding. Alternatively, start again from scratch, perhaps with weed-free turf.
- Moss can be discouraged by keeping shade to a minimum (perhaps by pruning overhanging branches), aerating the soil to encourage good drainage, and if the soil is very acid, using ground limestone to raise the pH.

up dose in spring, is usually effective. You can buy lawn sand or make your own from 1 part ferrous sulphate with 3 parts ammonium sulphate, and 20 parts sharp sand. Apply this at the rate of 4oz per sq yd (135g per sq m).

Moss killers containing chloroxuron seem to be particularly good at preventing spore germination (which means moss will re-colonize less quickly).

DISEASES

One lawn disease is very easy to recognize, the others need more skill – and probably a magnifying glass. Fairy rings (caused by several different fungi) are very distinctive. There is usually a circle of lush growth, and sometimes there is a ring of toadstools (especially in late summer or autumn). There may also be a narrow zone of dead grass. You can try aerating the soil, watering thoroughly, and using a strong solution of ferrous sulphate (sulphate of iron): 12oz in 1 gallon of water over 2sq yd (400g in 4.5 litres of water over 2sq m). On a high quality lawn the only answer may be to remove the turf to a radius about 1ft (30cm) beyond the ring, water with a two per cent formalin solution, then reseed or returf after four or five weeks.

Most other fungus diseases cause brown or discoloured patches (small in some cases, several feet across in others). With red thread disease you can see red or pink threadlike strands emerging from the grass. The use of a systemic fungicide (such as benomyl or thiophanate-methyl) and a feed at the right time will usually control most of them.

INSECTS

Irregular yellow or brown patches, especially noticeable in dry weather, may be caused by grubs, such as leatherjackets, cutworms, or sod webworms, feeding on the roots. Lots of birds pecking at the turf may be an indication of this kind of problem.

Use a soil insecticide such as diazinon. If you think that insect pests are going to be a problem, dusting the lawn with gamma-HCH in mid autumn is a sensible precaution.

REPAIRING A LAWN

Even the best lawns suffer from accidental damage occasionally, and an old, poorly made lawn may have bumps and hollows to be evened out, patches to be reseeded or returfed, and probably broken or damaged edges.

It's a wise strategy to get the grass into good shape first, then tackle repairs methodically during the course of a year. Any reseeding is best done in spring or early summer to give the new grass a season of growth without check; most other repairs that involve lifting or moving the turf are also best done in spring or summer to give the grass time to knit together again before the winter frosts arrive.

Lawn repair kits are available, but think carefully before you rush out and buy one. If it simply contains a small quantity of grass seed, and perhaps a piece of polythene with which to cover the patch while the seeds germinate, it may be cheaper simply to buy a small quantity of grass seed. Where the area to be treated is very small, however, a 'kit' may be a more economical way of buying a tiny quantity of seed. In either case it's essential to try to match the type of grass as closely as possible. It's no use reseeding a lawn consisting only of fine grasses with a mixture containing, say, a coarse ryegrass; equally a reseeded patch of fine grasses will stand out in a coarse play lawn.

Before reseeding, make sure the ground has been loosened, and water thoroughly with a gentle spray after sowing. Covering with a plastic sheet will keep

SODS AND TURVES

Sods and turves are the same, and which you choose to call them probably depends on which side of the Atlantic you live. In Britain 'turf' is the word usually used; in the USA it is 'sod'.

the birds off and help to keep the soil moist until the seeds germinate. Peg the plastic down with some pieces of bent wire, and be sure to remove it as soon as the seeds start to germinate.

Broken edges are quite easy to repair (see below) and, within limits, bumps and hollows can be levelled with a little determination and effort.

Shallow depressions can be filled gradually simply by adding about ½in (12mm) of sifted soil and peat, and repeating it once the grass has grown sufficiently, until the desired level has been reached. If the hollow is a large one, however, it will be necessary to make an H-shaped cut with an edging iron (or spade) over the area to be adjusted, then peel back the turf. Add sufficient soil and level it, then return the turf.

Bumps are also tackled by making an H-shaped cut in the grass and peeling back the turf. Only this time you will have to remove sufficient soil to level the ground.

In each case, water well if the weather is dry.

WHAT ABOUT WORMS?

Worms help to aerate the lawn, but if you want a first class ornamental lawn it may be necessary to discourage or kill them – they will mar the finish and the casts provide an ideal spot for weed seedlings to become established.

You can discourage worms by keeping the soil fairly acid (don't use alkaline fertilizers) and by not using topdressings high in organic material.

Wormkillers based on carbaryl should give a season's protection, and they should kill the worms below the surface (chlordane was once popular, but it is a toxic material best left to the professionals).

Derris dust can also be tried, but it brings the worms to the surface without necessarily killing them. Derris is not safe to use near ponds (it will harm the fish), but potassium permanganate is safe and also brings worms to the surface.

REPAIRING A BROKEN EDGE

1 A broken edge is a common form of damage, but fortunately one of the easiest to repair.

2 Use a straight edge and a spade (better still, a half-moon edger) to cut a rectangle in the turf.

3 Use a spade to undercut the rectangle of turf, carefully maintaining an even thickness of cut.

4 Lift and reverse the turf. It may be necessary to add or remove a little soil to keep it level.

5 Fill the gap, now on the inside, with soil, and brush soil into the joints. Reseed if necessary.

No matter how good the quality of the turf, it will look monotonous unless a large area is broken up by suitable focal points. Here a bird table serves the purpose.

Paving and Hard Surfaces

Paving (patios and paths) and other non-living surfaces come under the general heading of hard landscaping. It's worth bearing the term in mind, because it emphasizes the important role paving and paths play in garden design. And because it is so expensive and difficult to alter afterwards, it needs much more thought and pre-planning than planting flower beds or even making a lawn.

It's easy to start from the premise that paving is purely practical. Sometimes it is: you need paving around the house and a path to the front door simply to have somewhere clean and firm to walk on in wet weather; a paved area by the house might be regarded simply as somewhere to stand to hang the washing out, or for the children to play without getting muddy. But with good design, all these elements can be built into an integrated garden that blends the functional with the ornamental. This is especially so with paths.

PATIO AND TERRACE

One man's patch of paving is another man's patio. On a grander scale it might be called a terrace. This book is primarily about the choice and use of materials and surfaces rather than design, but it's impossible to divorce the two entirely, and it is helpful to bear a few design principles in mind when thinking about an area of paving.

A true patio should be regarded as a kind of outdoor room, and as such it needs to have 'walls' and 'furniture'. Low raised beds in matching materials (don't use block walling if the paving is brick or clay pavers; use a matching brick) give it boundaries and a sense of purpose. Areas left for planting small trees or shrubs produce an effect similar to container planting, without the watering problems. Planting areas left by the base of the wall enable climbers to be grown against the house. Screens of bamboo filter the wind and provide privacy. An area with overhead beams or trellis offers summer shade and a sense of enclosed security. A built-in barbecue makes the patio look more 'designed' than a portable one, but the materials used should blend in with the paving and any walling used, so that the whole area looks integrated.

All these things need planning. Built-in barbecues, raised beds, planting areas, are all so much easier to incorporate if you build them at the same time as the paving is laid. And it is often easier to obtain matching materials then. Adding piecemeal later can be a problem if a particular type of brick or paving or walling block is no longer available.

If patio overheads are used, the supports are much easier to incorporate and secure before the paving is laid. They should usually be concreted into the

Left: A garden for the disabled, with raised beds and a flat, even surface. But through incorporating a few bricks into the design, the concrete surface looks far from monotonous.

ground, which is a problem once the paving has been laid, perhaps on a concrete base.

A patio or terrace should be linked with the garden and not a feature apart. Linking it by steps or paths of matching materials helps to integrate the various features.

Don't forget that the patio does not have to be adjacent to the house. If there is a sunny, sheltered position down the garden, this is the place to have it.

Bear in mind also that paving does not have to run parallel to the house. Try laying it at a 45° angle. Do this, and use more than one material (perhaps clay pavers with old railway sleepers/ties, or paving slabs and bricks), and you will have a paved area that is above the ordinary.

For a large paved area, it's best to use paving slabs that are also large, with a background and plants to give height in proportion. Here, overhead beams give the illusion of a more intimate and secluded area.

USING BRICK

Brick is one of the most sympathetic man-made materials for the garden, and if you can choose one that is similar to the house brick, assuming you are paving near the house, this is one of the best choices for integrating home and garden.

Of all the paving and walling materials for the garden, brick is unfortunately the one that can be the most disastrous if you buy the wrong type. Often the type of brick used for house building is totally unsuitable for garden use. Unless it is very hard and frost-proof the brick will start to crumble and flake, even disintegrate, if water penetrates and then freezes. This may sound surprising if you think that homes get wet when it rains, and they are also subjected to cold temperatures. The difference is that bricks used for paving and garden walls will become exposed to dampness on all sides, and the moisture soaks through the brick; bricks used for house walls are only wet from one side, and the damp seldom penetrates far (and warmth from indoors ensures that the bricks don't become filled with ice crystals). Always make sure the bricks that you choose are suitable for the job (see Buying Hints).

The advantage of bricks is their versatility and subtle colouring. It's possible to use them for patios, walls, and paths, and they have strong yet subtle colours that do not fade. Because individual units are small, it is relatively easy to adjust to changes of level, and even lay them on a curve. Changes of texture can be achieved by varying the bonding pattern.

If you are not bothered about building matching walls, when bricks are the sensible choice, you can use clay pavers in a similar way to bricks and they have the advantage of being easier to lay (see page 62).

Pavers are thinner than bricks, but just as

A terrace gives the home a touch of elegance, and provides a natural transition between house and garden. The ballustrade and shallow steps down to the grass add distinction.

A narrow brick path assumes a sense of importance and considered design by the use of a herringbone pattern within the strong line of the edging bricks.

important is the slight difference in length and width. Pavers are designed to fit together without mortar; bricks have a mortar allowance, which means that they cannot be laid snugly together without gaps. They can be laid without mortar joints, but weeds will soon find a foothold.

Laying Bricks

A few kinds of bricks have holes running right through from back to front. These are not suitable for paving as they have to be laid on edge, which involves more cost and effort. But don't worry about bricks with a 'frog' (the name given to the recessed area in the centre of the brick on one side). Just lay them with the frog downwards.

Ground preparation depends on the wear and stresses that the paving is likely to have to endure. For light use it should be adequate to firm the ground and perhaps lay an 3in (8cm) layer of consolidated hardcore filled and levelled with ballast (sand and gravel) to provide a level surface. The important point is to ram and roll the surface level (albeit with a slight slope away from the house, to facilitate drainage).

Lay the bricks on a consolidated layer of sand about 1in (25mm) thick; if they have to take a lot of wear or weight, bed them in mortar.

Paving that has to take considerable weight or stress, a drive for instance, will require a proper concrete base, and bedding on mortar.

The weakest part of an area of brick paving is likely to be the edges. Try setting a row of bricks on end as an edging, securing them with a fillet of mortar.

Although a slight unevenness is less obvious with bricks than with paving blocks, uneven and poorly laid brick paving will be a constant source of irritation. Use a straight-edge and spirit-level

THE RIGHT FOUNDATION

A path or patio is only as good as its foundation, and the best materials poorly laid on a bad foundation will be uneven, and potentially dangerous if the paving becomes so uneven that someone may trip.

There can be no absolutes when it comes to recommending foundations, so much depends on the type of soil and the possible severity of frost penetration in winter. The following are typical recommended foundations for Britain; in very cold or much warmer areas, and particularly difficult soils, in the USA, they may require modification. In the States, boiling water or ice may be used for mortar and concrete in very cold or very warm weather. Even the sand may be heated, and special insulation and heating may be necessary if you are using concrete in very cold weather. Fortunately these extreme measures, and the additives used to overcome the limitations of climate, are unnecessary in the moderate climates of Britain and much of the USA. In very cold areas, frost heave can be a problem, and foundations have to suit local conditions. Even so, it makes sense to avoid laying concrete in the coldest winter months. You are not working to a construction site schedule, and delaying a project for a couple of months can do little harm.

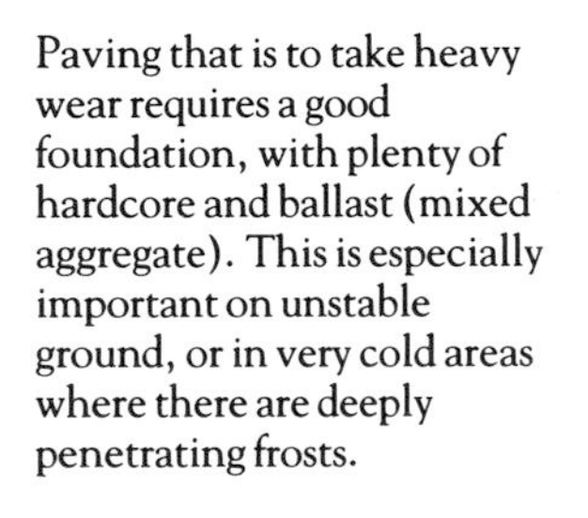

Paving that is to take heavy wear requires a good foundation, with plenty of hardcore and ballast (mixed aggregate). This is especially important on unstable ground, or in very cold areas where there are deeply penetrating frosts.

Foundations for Paths and Patios

Asphalt Use a firm surface such as gravel, ash, or concrete.

Frostproof tiles Lay on a concrete base at least 2in (50mm) thick (see *in-situ* concrete for foundations).

Gravel 4in (100mm) of well-compacted hardcore.

In-situ concrete Firm, well-compacted ground will be sufficient, but if there are unfirm areas use hardcore or ballast to provide a firm base.

Pavers For a path, well-compacted ground; for a patio it is best to use up to 4in (100mm) of compacted hardcore too. Top with 2½in (60mm) of bedding sand.

Paving slabs Well-compacted ground, if it is firm; otherwise 4in (100mm) of well-compacted hardcore.

Foundations for a Drive

Asphalt Use a firm surface such as gravel or concrete.

Gravel 4in (100mm) of well-compacted hardcore.

In-situ concrete A 4in (100mm) layer of well-compacted hardcore, levelled with ballast or aggregate.

Pavers 4in (100mm) of well-compacted hardcore. Top with 2½in (60mm) of bedding sand.

Paving slabs 5in (125mm) of well-compacted hardcore, levelled with sand or ballast (aggregate with sand).

constantly as each row of bricks is laid. Tap the brick into place with the handle of a club hammer or a mallet, and pack more sand or mortar beneath them if necessary. A garden line is invaluable for ensuring that rows are straight.

Pointing provides the finishing touch and professional effect that makes brick paving look smart, rather than becoming a catchment area for weeds.

A dry mortar mix can be brushed into the spaces between the bricks and water applied from a watering-can, so that it seeps through the mix and sets it. Use a piece of wood the width of the joints to compress the mortar before watering. You must take care to brush all traces off the surface of the bricks before the mortar dries, and it may be necessary to go over some parts

BRICK PATTERNS

Four popular and useful bonds for bricks or pavers. The stretcher bond is useful for small areas, and is easy to lay, but one of the other patterns will probably provide more visual interest for a large area.

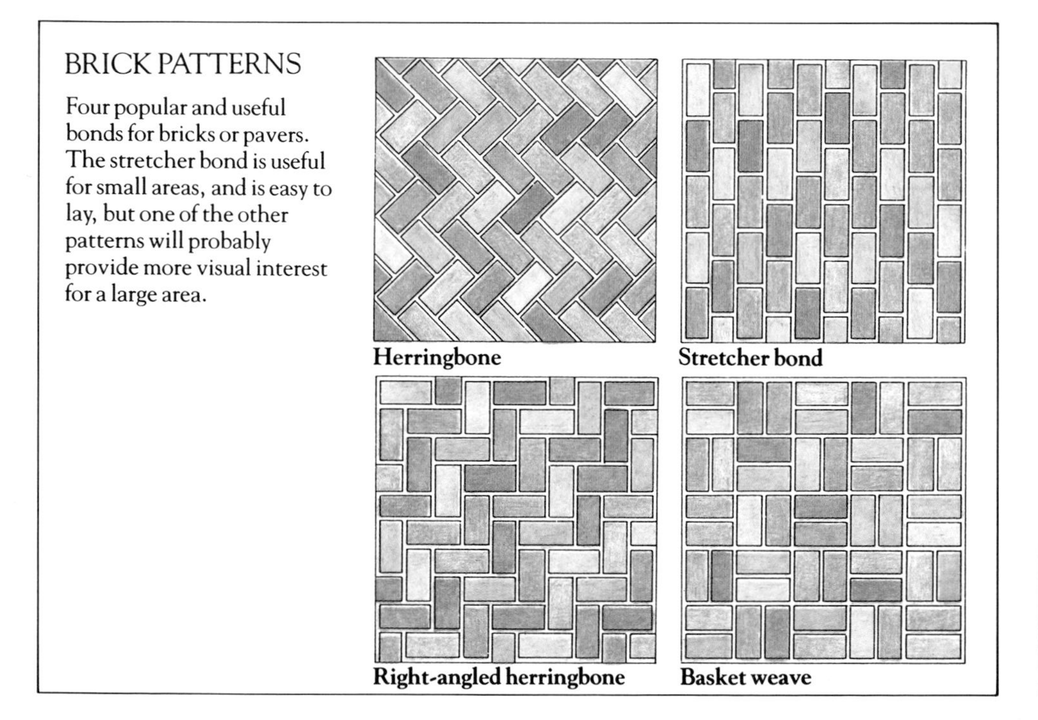

BUYING HINTS

Always buy bricks from a supplier who is a specialist, as you need to be reassured that they are suitable for garden work. Try to check with the manufacturer if in doubt. If you are buying a large quantity of bricks (say half a truck load), you may be able to buy direct from the brickworks. You will have a carriage bill, but this should be outweighed by the reduction in the price of the bricks.

Generally soft 'facing' bricks, used for houses, are unsuitable. In Britain, those described as 'special quality' should be satisfactory. Very hard engineering bricks are strong enough, but not particularly attractive for a garden setting.

It is also best to avoid bricks with a very smooth surface, as they may become too slippery when wet. Some bricks have a very rough finish that can look charming but may not be a good choice if children are likely to fall and get grazed knees.

Other terms you may come across: **calcium-silicate** bricks are a special kind with a more 'regular' appearance than ordinary bricks, and generally a pale, even colour – some are suitable for garden use; **flettons** are bricks made from a particular type of clay; much used for house construction in Britain, but not suitable for the garden; **stock** bricks have a relatively low density and strength, and generally a rather rough and 'rustic' appearance **('rough stock'** bricks are seconds that are a little overbaked and misshapen, not a drawback for the garden); **wire-cut** bricks have a more regular profile and finish than, say, stock bricks. Unless they have been given a finishing texture, they are often smoother than other bricks.

Right: A brick path or patio looks best with an edging of bricks laid at a different angle. A line of bricks flatly laid at right angles to the main pattern neatens the shape and gives a stronger finish. The edging can provide an extra design element: in the path shown below, the bricks are set upright and at an angle to give a raised edge.

with a moist mortar mix afterwards to fill in any gaps. You can fill the cracks completely with a trowel, but it can become tedious for a large area, and it's still difficult to keep mortar off the surface of the bricks.

Before the mortar sets, run along the joints with a piece of dowel, or some other rounded implement of a suitable size, to produce a smart, slightly recessed finish. This generally looks better than a flush finish, which detracts from the crisp outline of the individual bricks.

The choice of laying pattern will have a big impact on the overall impression of the area. Some need to be seen over a large area for the effect to be striking. Others, such as basket weave, are successful even for a small area.

LAYING A BRICK PATH

Prepare the site for the path by removing sufficient soil to take about 3in (8cm) of hardcore and 2in (5cm) of mortar, plus the depth of the brick. Sand can be used instead of mortar, but make sure it is well compacted.

Start by laying the edging bricks along one side of the path, banking mortar on the outside to hold them firmly in place. As you proceed with laying the flat area of the path, set the edging bricks in place in the same way along the other side.

1 Prepare a 2in (5cm) layer of mortar mix (1 part cement to 4 parts sand), and bed the bricks on this. Tamp the bricks into place, using a straight-edge and spirit-level.

2 Brush a dry mortar mix between the joints, making sure there are no large air pockets (use a piece of wood the thickness of the joints to compact it if necessary). The mix as used for the base is suitable.

3 Once the joints have been filled with mortar, sprinkle water from a fine-rosed watering-can. The fine mist from a compression sprayer is even better. Clean any mortar off the surface of the path before it dries.

USING PAVERS

Unless you are trying to match brickwork to garden or house walls, pavers can be a much better choice than bricks. There are clay pavers that look like bricks (they are thinner and a slightly different size so that they fit together tightly without mortar joints), and concrete pavers that lack the warm appearance of the clay versions but are nevertheless very useful.

There's the merit of being able to lay pavers on a sand base, with no need for mortar. And you need have no doubt about their strength despite this – the same pavers are used, with the same basic principles of laying, to provide roads and standing areas suitable for buses and heavy vehicles.

BUYING HINTS

Pavers are sometimes bought by the square yard/metre, though they are also sold individually. You should be able to calculate the amount fairly accurately (on graph paper if necessary). Allow up to 5 per cent for breakages (you will probably get away with less if you don't have to cut many). The difficulty comes in estimating edge pieces for blocks that are not rectangles, but the manufacturer should be able to provide the formula for working out how many you need (or ask your supplier's advice). The vital measurements that you will need are total area (length × breadth), the linear yards/metres of the ends, and the linear yards/metres of the edges (sides).

Left: A clever mix of natural and man-made materials, again with a useful change of level to add interest.

LAYING PAVERS

Prepare a base of compacted hardcore and aggregate (see page 59), and use curbstones or edging blocks to prevent the paver blocks moving. Make sure the final height will not be above the damp-proof course. Allow a fall of 2in (50mm) in 6ft (2m) for drainage.

Spread sand over the area, and use a straight board to level and compact the sand. Notch each end of the board to the right depth, and tamp between two boards at the final height

Simply lay the pavers in position, butting each one firmly against its neighbours. Start in one corner and then work along the main straight edge. Work with full-sized blocks as much as possible (below left).

Once an area about 12ft × 12ft (4m × 4m) has been laid, compact the blocks – ideally with a flat-plate vibrator (which you can hire – but ask for one with a rubber sole plate so that the paving is not marked). Don't take it near an unsupported edge. Once all the pavers are laid, brush sand into the joints, and go over the area with the vibrator again to lock the blocks together. Get a helper to brush the sand in front of the vibrator as you use it.

Most clay pavers are simple rectangles, like bricks, but there are also shaped ones that interlock like a jigsaw. The concrete type also come in rectangles, but are much more often seen as interlocking blocks.

This type of 'flexible paving' is quick and easy to lay, but you must have firm edges. Mortar the edging blocks into place to resist the possibility of sideways pressure from the finished paving. In the case of interlocking blocks or pavers, you will need special edging pieces (in effect, the straight edges of the jigsaw).

Don't be put off by the thought of concrete blocks. Apart from the usual grey, they come in shades such as browns, and because the finished paving is a pattern of small units, it will not look bleak or uninteresting. They actually blend in well with many garden situations. Clay pavers the approximate shape of bricks can, of course, be laid to all the patterns suggested for bricks (see page 60). Although any pattern is suitable for foot traffic, for a drive it is best to use a 45° or 90° herringbone pattern.

Cutting paver blocks

If many pavers have to be cut, it is well worth hiring a hydraulic block-splitter – it is easier and more accurate than using a hammer and bolster. If you have to split blocks or pavers by hand, score the cut line all around, and strike a bolster chisel firmly with a club hammer to make a clean cut. A firm, level surface, perhaps a bed of sand, is necessary for a clean cut.

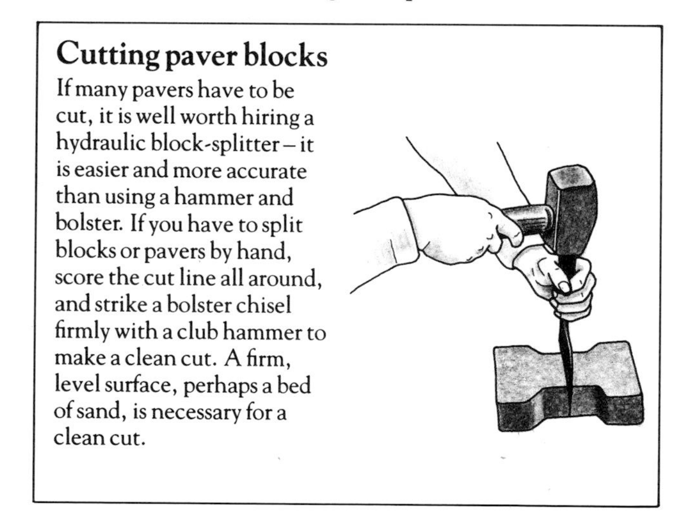

USING PAVING SLABS

Paving slabs can be mortared onto a concrete base, and this has some merit around the side of a house, or for a drive. But for a patio or path, it's better to spot mortar onto sand over a hardcore base. This will help if you ever have to lift a slab, and it makes the whole job easier and less expensive.

For light or moderate use (perhaps wheeling a heavy wheelbarrow), 2in (50mm) of consolidated hardcore covered with 1in (25mm) of raked and compacted sand will provide an adequate base. There is no point in making foundations deeper or stronger than actually necessary for the job. (If firmer foundations are necessary, see page 59.) Always ensure that the final level of the paving is below the damp-proof course.

Because paving slabs are larger than pavers and bricks, errors in alignment and level are much more obvious, and it is necessary to work to string lines to keep the rows straight.

Advance planning is essential, not only to work out the number of slabs required, but to know where to leave out any slabs for planting trees and shrubs. If you are using paving slabs of mixed sizes, the whole plan will have to be worked out on graph paper first. Bear in mind that using, say, three different paving slab sizes can make laying more troublesome and will not necessarily look any better than a simple design using slabs of just one or two sizes.

Be particularly cautious about colours. Some makes of paving slab can be rather garish when new and don't always look tasteful. By the time they have weathered for a few years the colours will be much more muted with the patina of time – so why not settle for fairly muted colours to start with?

Surface texture and finish are more important than colour. Exposed aggregates can be less slippery in wet weather, and can look less boring (although any of them can be over-used). Some have a rippled finish that resembles natural stone, and this can be attractive.

BUYING HINTS

Always allow a few slabs spare for breakages. If you have to cut a fair proportion of the slabs, allow up to 5 per cent for waste.

The thickness of paving slabs can vary, larger sizes sometimes being thicker than small slabs. If you are mixing sizes, make sure they are all manufactured to the same thickness.

Different manufacturers use their own names for some of their finishes, but the following terms are generally recognized:

cobbled slabs have a finish that resembles old cobble stones;
exposed aggregate slabs have small gravel in the concrete exposed at the surface;
riven slabs have a weather-worn appearance, with a slightly uneven surface;
stipple slabs have slightly raised spots or streaks.

You may find some described as cast and others as hydraulically pressed:

cast slabs are made in a metal tray mould; they suffer from the drawback of having joints that do not butt so well (but mortar joints will overcome this), and are more vulnerable to chipping;
hydraulically pressed slabs have been produced, as the name suggests, in a hydraulic press; they are suitable for either mortar or butt joints.

A small garden, with interest provided by the slight change of level and a raised pond.

LAYING PAVING SLABS

Prepare the foundations (see page 59).

Mark out the first row with a line, using a builder's square if necessary to make sure that the angles are correct. Start from the house, or a known straight edge.

There are various methods of fixing paving slabs, and a popular one is to place five fist-sized dabs of mortar where the slab is to be laid. A recommended method is to use a box and square of mortar (see illustration).

Place the slab over the mortar and tamp it level with a wooden mallet or the handle of a club hammer. Use a spirit-level on a straight-edge to make sure the slabs are even (use a shim of timber to allow for any drainage slope).

To ensure that the joints are of an even thickness, use pieces of wood of the right thickness as spacers between the slabs. Make sure you cut an adequate supply of these before you start (or buy suitable spacers).

Finish off with mortar rammed between the slabs. A recessed joint of about 1/8in (2mm) looks good, but deeper joints may catch heels. Use a wet sponge to remove any mortar on the surface.

Cutting slabs

To cut a paving slab, simply score a line carefully at the right position, *on both sides and edges of the slab,* then chip a groove along the length with a bolster chisel and club hammer. Once there is a clear groove, increase the force of the strokes until the slab splits. Alternatively, lay the slab on a wooden batten at this stage, and tap along the *back* of the line until the slab breaks. Try both ways and see which you find the most successful. Some types of slab can be cut much more easily than others, so it makes sense to work to a size that avoids the necessity for cut slabs whenever possible.

USING TILES

Quarry tiles (which are either brown or red, and come in many shapes and sizes) and frost-proof ceramic tiles are ideal for a patio that you want to appear linked with the house or conservatory. They have more of an 'indoor' look than the other forms of paving described, and can look very tasteful.

Not all ceramic floor tiles are suitable for outdoor use, and you must be sure that they are frost-proof otherwise they may crack in the winter. As you can't tell how frost-proof they are simply by looking at them, you must receive a firm assurance from your supplier that they are suitable for the job. Bear in

Right: Frostproof ceramic tiles – a refreshing change from concrete paving slabs. **Far right:** A clever contrast between the soft form of the plants and the regular outline of tiles with a dominant pattern.

LAYING TILES

The picture sequence below shows how to lay ceramic tiles, a relatively easy task once a firm and level foundation is in place.

Thicker quarry tiles (about 2in/50mm) can be laid on sand or over a concrete base of 1 part cement to 5 parts ballast. Use battens to contain the concrete and act as a guide for the finished tile level.

Soak quarry tiles for about 15 minutes before laying. Spread mortar over the concrete base and lay the drained tiles on the mortar. Tap them down flatly with a block of wood and use a straight-edge and spirit-level to check that the finish is even. After 24 hours you can point the joints with a runny mortar mix, but wait three days before walking on the tiled area.

1 A smooth, level concrete base is needed for laying ceramic tiles. The tiles are bedded on a special adhesive (which can be obtained from the tile supplier) applied with a spreader.

2 Use a combing tool to work over the adhesive layer and provide a 'tooth' for laying the tiles. Work a fairly small area at one time, so the adhesive does not start to dry before the tiles are laid.

3 Lay each tile on the adhesive layer cleanly butted to the adjoining tiles. When the tile is in position, press down firmly with both hands to bed it evenly.

mind also that glazed ceramic tiles can become very slippery after a light shower or when they are covered with dew.

Some tiles come in attractive shapes, but the rectangular tiles are easier to lay – especially when it comes to the edges. Cutting some of the curved or shaped ceramic tiles (which are hard and brittle) can be extremely frustrating and wasteful, even with a tile cutter. There's a lot to be said for laying the main part first, leaving the edges until the end, then marking all the cuts in pencil and taking the tiles to a stonemason with a diamond saw. This will produce a nice crisp and accurate edge with a lot fewer breakages! If you are using rectangular tiles, however, you may be able to cope yourself.

USING CRAZY-PAVING

Crazy-paving can be an inexpensive option, as broken paving slabs are cheap to buy. But broken coloured paving slabs can look very artificial and unsympathetic in many garden settings, and should be used with care. Laying can be a tedious, messy job, and getting a nice even finish is not easy.

Crazy-paving is at its best when a natural stone is used, especially if planting holes are left to accommodate suitable small plants. But as natural stone is likely to be expensive you lose the financial benefit.

If you have an uncompromising concrete or paving slab path that you want to improve, topping it with natural stone crazy-paving, such as slate, can do much to enhance the garden without the labour of digging up the original path and laying new foundations.

Opposite: Natural stone is particularly suitable for crazy-paving – here slate has been used.

LAYING CRAZY-PAVING

Prepare a foundation of compacted hardcore if there is not an existing path or foundation. Lay the paving stones or slates roughly as you expect to use them (below left), keeping straight-edged shapes to the outside. Lay large pieces first and fill in with smaller ones, trying to make fairly even joints. Thick stones may have to be split.

Bed the slabs on a mortar mix of 1 part cement to 3 parts sand, packing extra mortar beneath uneven stones to bring them level. Check frequently with a straight-edge and spirit-level as you work (below right).

Fill the cracks with mortar (coloured to match the stones if possible), leaving crevices for growing plants if required. Wipe off surplus mortar while it is still moist and use an old knife to clean off any drying mortar on the edges of the stones.

ASPHALT FOR GARDEN USE

Asphalt is a practical surface rather than a pretty one, and best confined to drives and formal paths in a utilitarian part of the garden, such as paths to the garden shed, around the kitchen garden, and similar situations. An asphalt finish doesn't have to be black – you can buy it coloured red or green. And stone chippings enhance the surface texture.

A large drive is perhaps best done professionally. The cold macadam that you buy for DIY work can become more expensive if you have to cover a large area. However, it is easy to lay provided you have a roller.

You can lay asphalt on any firm surface, such as gravel or concrete. Level and prepare the surface first, allowing a slight fall for drainage as necessary. Concrete edgers provide a neat finish to the asphalted area.

Apply a layer of bitumen emulsion (tar paint) to bind the surface. When this has turned black, pour the macadam from the bag and rake it out to a thickness of about ¾in (18mm). Level and firm it carefully before compressing with a roller, dampened to prevent the macadam sticking to it. Top up any depressions that appear in the rolled surface.

If stone chippings are to be used for the surface finish, sprinkle them over the surface and re-roll to bed them in.

Coloured asphalt can make a strikingly bold complement to the grassy carpet of the lawn. In this garden a double path of red asphalt forms a clean-edged driveway running through the rather formal design of the planted areas surrounding a modern cottage-style house.

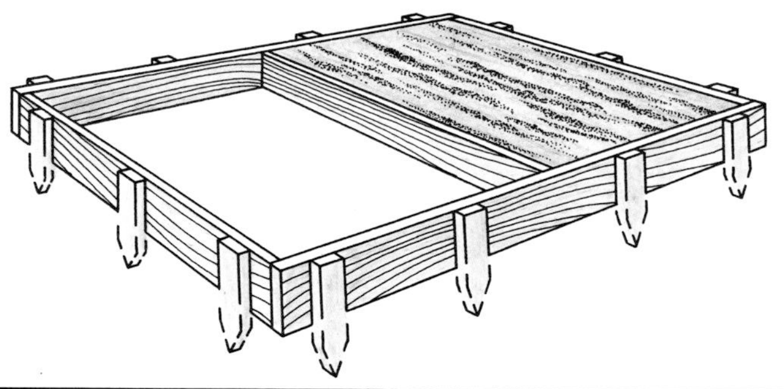

Concrete paths needn't be boring. The combination of inset timber and a textured surface produces an interesting finish. Concrete should be laid in small sections defined by timber formwork (right).

USING CONCRETE

Concrete is frequently regarded as a second-rate surface, at least aesthetically. That's probably because it's usually associated with concrete drives finished at best with a slightly rippled surface.

Professional designers sometimes use concrete very imaginatively, and by taking a few tips from them it's possible to provide a very practical surface that is no cause for embarrassment in the context of garden design.

There are three ways in which you can enhance concrete: by texturing the surface, colouring it, and using inserts of other materials (from railway sleepers/ties to beach pebbles).

Concrete needs to be laid carefully, following certain rules, if it is to remain the tough and trouble-free surface that it can be. Prepare the foundations to suit the use. Excavate to the required depth, cut back any tree roots, and be careful to avoid any buried services (electricity, water, gas, drains); don't leave these any less exposed than they were before. The sub-base of hardcore should be 4in (100mm) thick, and the thickness of the concrete slab at least 4in (100mm) if cars and light vans are to use it (make it 6in/150mm on soft soil or on clay). For ordinary garden paths, you may not even need a sub-base except on unstable soils (such as peat or clay), and 3in (75mm) of concrete will be adequate.

Unless you provide expansion joints the concrete will almost certainly crack later. The joints should always meet the edge of the slab or path at right angles if possible; it's a good idea to have these about every 3ft (1m). You can use a treated wood (redwood has natural rot resistance) for these, and leave them in place as part of the design. But if you are mixing materials, you may be able to work in railway sleepers/

ties, even a row of bricks, at the point of the expansion joint. Try to avoid intersecting both materials and expansion joints, otherwise it will look too fussy.

Don't forget that a wide expanse of concrete will require expansion joints along its width too – avoid laying a solid width greater than about 10ft (3m). Much above this, it's best to lay the slab in two parallel equal-width strips.

Mixing concrete yourself is very hard work. Wherever it is practical (especially if access is good), it pays to get ready-mixed concrete delivered. If you have a large area to lay you may need an army of helpers, to complete the job in the 'workable' time, but it's worth the effort to get it done this way. If you make it clear what the concrete is required for, you should be sure of getting a suitable mix, with entrained air. You will have to have all the formwork and expansion joints already constructed in position.

If you are dealing with a small area, and doing it yourself, use a mix of 1 part cement, 1½ parts sand, and 2½ parts 20-mm aggregate (or alternatively 3½ parts combined aggregate).

Right: To avoid concrete paving slabs giving a dull effect, try working in some large pebbles between the slabs. Use a straight-edge to ensure that they are set level with the paving, so that they are not uncomfortable to walk over.

Surface Finishes

The previous page shows how formwork is used and concrete laid. It is in the final finish that you can provide the ordinary with a touch of character.

Exposed aggregate is non-slip and generally more interesting than a smooth finish. Spread some coarse aggregate evenly on the fresh concrete and tamp it gently into the surface. Once the concrete has hardened enough for the aggregate not to be dislodged by the action, brush the surface and spray with water to wash away fine particles. This will leave the stones standing proud of the surface. A very fine gravel can be used for a fine textured surface.

Brushed finishes are simple to create and varied in appearance. A soft broom will produce a fairly smooth finish (use it on fresh concrete after the smoothing pass with the tamping beam). A stiff broom will leave a 'corduroy' effect if you trail it across the surface with the bristles at a fairly shallow angle.

You don't have to use a brushed finish for the whole area – try alternate 'bays' between expansion joints, or make the brush marks in different directions in each section. The marks don't have to be in straight lines, either; swirls and circles make an interesting variation.

Pebbles can be pushed into the surface and levelled by using a block of wood over the tops to tamp them. Crushed stone can also be tamped into the surface. Brush and wash the pebbles or exposed stones before any slurry on the surface sets too hard.

Stamped patterns can be used to create an effect

similar to bricks (they create several 'courses' at a time; ideally the slab of concrete needs to be in a multiple of the stamping tool area). You may find it difficult to obtain these tools, but you can improvise with all kinds of other impressions. Pastry cutters and sea shells are among the 'tools' that can be used for leaving an impression – but give it all a lot of thought first otherwise you may regret it months later if it looks too frivolous.

Leaf impressions can look interesting, just as fossils in rocks provide a talking point. Use leaves with a distinctive shape or texture, such as sycamore or horse chestnut, and press them into the surface with a trowel; once the concrete is hard, brush the leaves out to leave the impression.

Coloured concrete can be effective, but try a small area first. The colour can look different once the concrete has dried. It needs foresight and experience to get good results with coloured concrete. Not only has the final colour to look right, but you need to mix exactly the same amount of pigment for each batch, and if you put too much or too little water in any batch there will be an uneven intensity of colour (more noticeable with 'strong' colours).

INSPECTION COVERS

Inspection covers in an area that you want to pave can be a real problem, but you can buy special covers into which you can inset shallow pavers or bricks.

Drain and other services inspection covers are difficult to deal with if they occur in an area that you want to pave. If you are laying concrete, it will have to be flush with the inspection cover, which will have to be shuttered with timber when the concrete is poured.

Where bricks, pavers, or slabs are used, an inspection cover may be an unacceptable visual intrusion. Fortunately you can buy trays that fit into standard inspection covers, with a recess deep enough to take clay pavers, for instance. You will probably have to cut some of the pavers to fit, and it's best to use an angle-grinder (or get a professional to cut them for you) for a neat finish.

There are also shallow planters that are designed to fit into an inspection cover, and if it's in a part of the patio where a small patch of flowers would be appropriate, this is well worth considering.

Another option for, say, crazy-paving or clay pavers, if you can't find a proprietary cover, is to lay the paving over that area on a sheet of butyl rubber (the kind used for pond liners). Mortar the pieces as normal, but make sure the mortar and paving is all contained within the butyl and can be lifted if necessary. This is only suitable for inspection covers that are not likely to be used often, as you will have to remortar again after the sheet has been lifted for inspection.

If possible, avoid standing containers on inspection covers. Unless they are larger than the cover (in which case they will probably be too heavy to move easily), they will draw attention to the feature rather than hide it.

All inspection covers should be easily accessible when necessary.

DECKING

Timber decking is seldom used by amateurs in Britain, but in the USA and parts of Europe it is much more popular. Although not as permanent and maintenance-free as, say, concrete or brick paving, timber blends with the garden setting especially well. And it doesn't look as harsh as some of the alternative paving materials for a patio. Timber decking is also a very effective way to achieve a link between home and garden, especially if the house has timber cladding. It is not so useful in damp climates, as wood absorbs moisture and can cause expansion problems (and dampness speeds up the rotting process).

Grain, texture, and colour all make wood a very exciting material to use. Different patterns and spacings, different angles, and various kinds of edgings and railings, all give timber decking lots of design potential.

The right choice of timber is essential, however, if disappointment is not to follow in just a few years. Some woods have a natural rot resistance (redwoods for instance), otherwise it is essential to use timber that has been pressure-impregnated with a suitable preservative. The actual timber available will depend on where you live: those in Britain may not be the same as in the USA, where availability will again vary from one part to another. It's best to go to a good timber merchant and seek his advice (and you'll get accurate prices for comparison too).

If you are not aware of the weathering characteristics of the different timbers, ask to see a piece of weathered wood – it is likely to look very different from the new timber. If you are inspired to try timber decking, it is worth reading more specialist books on the subject as there is not space in a general book of this kind to deal in detail with structural aspects like beam spans and joist spans, which you need to take account of for a strong structure. You do not want your deck to flex or bounce so that it feels like a diving board when you walk along it.

DECKING CONSTRUCTION
A chequerboard effect (right) is obtained by laying the timbers to form squares with the timber lengths set in alternating directions. The wooden framework for the decking construction stands on a layer of gravel or sand over a hardcore base layer (below).

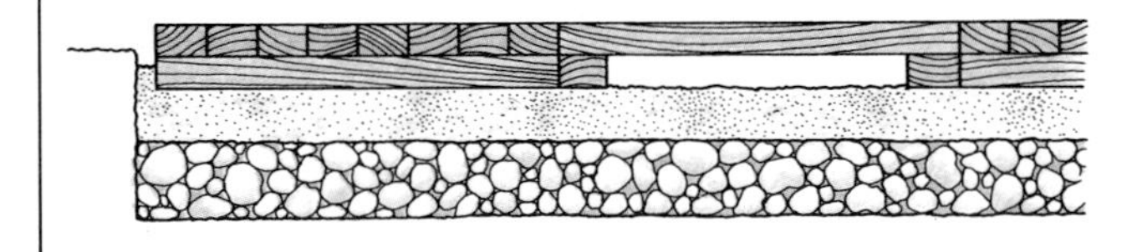

And if you plan to have heavy furniture or unusual loadings on the decking, you will have to make additional allowances. If you buy 'kit' decking, it should be structurally sound and made of suitable timber.

Bear in mind that even ground-level decking will need a firm foundation, such as a concrete slab (you may be able to buy concrete supports on which you can place the beams), and for raised decks proper footings are required to support the structure.

As a general guide, the decking timber should not be more than 6in (15cm) wide, and for light general use 2in ×6in (50mm ×150mm) or 2in ×4in (50mm×100mm) timber will span 4ft (1.2m) if laid on its side; if the wood is only 1in (25mm) thick, the maximum span should be reduced to 2ft (60cm).

Timber paths can be attractive, and you can buy modules from which to build up a design. These will have been treated to make them durable. Modules may be 2ft (60cm) wide, so by doubling up you can easily make a 4ft (1.2m) path. By alternating the direction of the timber in the two modules that are parallel, the whole path can be given a more interesting appearance. Lay the modules on sand over a base of gravel for good drainage and to provide a firm and level surface.

For any timber project in the garden, it's best to use pre-treated timber. The kind of preservative that you paint on yourself will not give enough protection for decking or a timber path. Not all pre-treated timber is equally suitable for all garden uses. In the USA you will probably find the timber marked as suitable for above-ground use, ground contact, and foundation use. Without such information you will have to ask.

Left: Even a much-used play area can be attractive, if it's well designed, as this picture shows.

You will have to shop around for coloured gravels, but they are available (sometimes a garden centre will be able to order them for you, otherwise try a stone merchant or a good builder's merchant). Here coloured gravels have been used in a formal setting.

GRAVEL

Gravel has lots of potential for the garden floor. It's cheap, easy to maintain, and visually attractive. It's also very versatile and can be laid into any shape, no matter what the curve or angle. And it blends well with other garden floor materials, such as paving flags, natural stone, and bricks. You can also plant through it to produce beds of carpeting plants or even dwarf shrubs, wherever they seem appropriate.

There are many different kinds of gravel, from small 'pea' gravel to large angular kinds that have a much coarser appearance. And colour varies too, according to source. The appearance of gravel also changes with the light and whether or not it is wet: the same gravel may look bright and almost glaring when it is dry and in full sun, yet brown and muted and very restful when wet or in the shade.

There are three main uses for gravels: as a simple lawn substitute, where it is used to fill the space between fixed edgings; as a natural textured surface through which to grow some plants informally; as an infill among other forms of paving. Gravel can also be used for paths which, though difficult for wheeling barrows and possibly noisy underfoot, have a natural charm that more ridged materials lack: and of course gravel is ideal for a meandering path that would be difficult to lay with fixed-size modules like paving slabs.

There are a few drawbacks with gravel. If the gravel is above the level of the surrounding ground, it will easily be scattered (you need a firm restraining edge for a gravelled area). And you can't walk on it with bare feet – you will also find it difficult to use with high-heeled shoes.

Weeds are sometimes mentioned as a potential problem, but no gravel area need be marred by weeds. For general areas, lay a sheet of heavy-gauge plastic sheeting over the ground, and put about 2in (5cm) of gravel on top of this. The plastic will eventually disintegrate, but it will avoid perennial weeds growing for several years. New seedlings find it quite difficult to become established, and if necessary can be hoed off relatively easily. If weeds do start to become a problem, and you don't mind using garden chemicals, a path weedkiller will easily keep gravel weed-free for a season.

Gravel paths and drives require a little more preparation. Provide a firm base by excavating the ground to take about 4in (100mm) of compacted hardcore and 3in (75mm) of gravel for a drive or 2in (50mm) for a path. Concrete edging can be used for a formal path or a drive, but treated timber will be adequate for most paths. Hoggin can be used instead of hardcore. Rake and roll the gravel to produce a slight camber for drainage (watering it occasionally may help compaction). Apply the gravel to half the depth first, then finish off to the final level.

BONDED GRAVEL

For a small decorative area of a patio, perhaps mixed with paving, or as a pool surround, it may be worth considering a bonded gravel. This can look effective for a path too, in conjunction perhaps with a row of paving slabs along the centre. It makes a fairly smooth, attractive finish, giving the visual effect of fine gravel without any loose pieces.

Bonded gravel is sold in kits, containing the small pea-type gravel and the resins that you mix to bond the gravel together and bind it to the prepared surface. The kit will provide all the necessary instructions.

GROUND COVER

Ground cover is something that professional gardeners and garden designers realize the value of much more than amateurs, yet it can be as relevant for a small town garden as for a country estate or a public park.

The benefits are clear. Ground cover plants smother weeds and reduce or even eliminate the need to weed. They cover areas of ground that would otherwise have to be maintained more intensively (as lawns, shrub borders, herbaceous borders, and so on), and they add 'texture' to a garden with drifts or blocks of plants creating an overall impression of design, linking paths, lawns, and beds in a way that is difficult to achieve with mixed plants. Some will clothe the ground in inhospitable places such as the area beneath trees, which would otherwise remain largely bare ground.

In some situations, ground cover plants play a crucial role in preventing soil erosion, and in a few cases, such as very steep banks which would be difficult to mow if grassed, make it possible to clothe the soil attractively in a way that would be almost impossible with mixed plants.

Little is achieved without effort, however, and all these benefits accrue only if the ground has been well prepared and carefully maintained during establishment. Getting ground cover established can be more work than looking after a lawn – and it may be required for weeks or years, according to the plants used.

Being realistic and choosing *suitable* ground cover is the other key to success. A plant that is not happy, not naturally adapted to the conditions provided, will not make the kind of dense growth and good ground cover required. On pages 91 to 117, there are selections of plants suited to particular soils and conditions (sun or shade, for example). Use these as a guide, and don't be influenced simply by the appeal of, say, the flowers. Ground cover can be very decorative, but it also has to be practical and successful. That's its *raison-d'être.*

Speed of cover is partly a matter of growth habit, but also planting density and growing conditions. Close planting achieves results a season or two sooner, but it could double the cost. And as the slower growers are often the more expensive plants, the financial implications of using some of the choice but slow-growing kinds are significant.

— DESIGNING WITH GROUND COVER —

It's possible, and perfectly satisfactory, to regard ground cover plants simply as a means of clothing a problem area of ground, with the minimum of maintenance. Yet the effect will be that much better if ground cover is seen as a design element too.

Left: Heathers are among the most beautiful and efficient flowering ground covers. This is *Erica* × *darleyensis* 'A. T. Johnson' in mid spring.

Hostas provide beautiful ground cover but, of course, they die down for the winter. Fortunately the dense summer foliage discourages annual weed growth if the ground has been cleared well initially.

Ground cover is at its most striking in large drifts or blocks, but not so large that they become boring. Don't be afraid to break up a very large area with smaller blocks of different colours or textures, especially effective when different varieties of the same plant are used. A very large area of the plain-leaved *Pachysandra terminalis* will look more striking with a broad band or snake of the white-edged 'Variegata' going through it. A carpet of a green-leaved ivy will be more eye-catching with a contrasting patch of a variegated form.

Where ground cover is used largely as a substitute for grass, hard paving needs to be carefully placed so that there is the visual effect of a lawn without the necessity to walk on the planted area. The danger lies in making rigid paths that dissect the area with too many straight and uncompromising lines – the effect you get when a straight path and perhaps washing line dissect a grass lawn in a small garden. Make a point of using paths diagonally, or perhaps zig-zagging across the area (provided this does not encourage short cuts across the plants). For a little-used path, stepping stones can be very effective among ground cover, much more so than in a lawn, where they often look too contrived.

Try to integrate ground cover with hedges. Some ground cover plants, such as ivy, will grow happily at the foot of a hedge, which is something few flowering plants will do readily because of the competition for light, moisture, and nutrients.

An ivy, or perhaps *Walsteinia ternata*, will smartly fill a narrow strip of ground between a hedge and a path – and if treated as a texture or pattern in a montage of surfaces, the juxtaposition of hedges, paths and ground cover can give a garden a smart, designed look. It is sometimes a better treatment for small modern gardens than a more natural and informal style that is very difficult to achieve without the space and background that help so much.

Don't overlook colour. Try the variegated white-edged *Pachysandra terminalis* 'Variegata' in front of a purple-leaved hedge such as *Berberis thunbergii* 'Atropurpurea', a yellow-flowered genista or *Hypericum calycinum* in front of a green hedge.

A light or brightly variegated plant is also a good choice beneath trees, in an area that often looks

Sun roses (*Helianthemum*) on a sunny bank. These brightly flowering plants must have a sunlit position and will tolerate dry soil.

Left: *Pachysandra terminalis*, one of the very best ground covers for a shady place. Note the textured finish of the surrounding concrete.

gloomy. A carpet of the large-leaved variegated Canary Island ivy (*Hedera canariensis* 'Variegata') or a gold-splashed *Euonymus fortunei* 'Emerald 'n' Gold' at the front of a shrub border, will bring a touch of lightness even to an area of shade or partial shade.

Although more suitable for a large garden than a small town plot, the whole garden can be devoted to ground cover interesected by paths. It needn't look formal if you choose narrow meandering paths, perhaps including stepping stones in parts, picking their way through the various carpeters.

This approach depends on having a range of different ground covers to provide sufficient variety and interest, and a few dwarf shrubs and even small trees (such as Japanese maples) to provide height. Mixing the ground cover plants can present problems, because of their different growth rates and habits, but they can be separated effectively by even a narrow path. But from a design viewpoint it is not sensible to have paths, no matter how attractive, separating all the different ground cover plants, and it will add interest if as many as two, three or even four can be used in one bed provided the drifts are bold enough for impact. To prevent one growing into another, simply maintain a 'river' or channel of uncultivated soil between adjoining species – it does not have to be more than about 6-8in (15-20cm) wide and a dressing of gravel or pulverized bark will look more attractive than bare soil. A carefully applied path weedkiller will keep these dividing ribbons weed-free for a season, and an occasional trim with shears or edging shears (as used for lawn edges) will prevent shoots encroaching from either side.

This type of gardening gives the impression of a plantsman's garden, yet with real low-maintenance benefits. The whole effect is destroyed, however, if weeds become established before the ground cover, so

ADJUSTING THE pH

Whether you are growing ground cover plants like the ones in this chapter, or a conventional lawn, the pH of your soil can affect how well some of the plants will grow. In the case of a lawn, the pH (which is a measure of acidity or alkalinity) can influence things like worm population, moss growth, and even the types of weeds that will be a problem. Simple kits can be used to determine the pH of your soil.

Generally, it's best to grow plants that suit the soil, rather than trying to make it more acid or alkaline. But if you do want to make a modest adjustment, the following amounts of lime should decrease the alkalinity by about 1pH. It's best to apply it in several small doses rather than all at once.

Soil	*Hydrated lime*	*Ground limestone*
Clay	18 oz/sqyd (425g/sq m)	24oz/sq yd (575 g/sq m)
Typical loam	12oz/sq yd (285g/sq m)	16oz/sq yd (380g/sq m)
Sand	6oz/sq yd (140g/sq m)	8oz/sq yd 190g/sq m

If you want to make your soil more acid, the following should increase the acidity by about 1pH, but bear in mind that sulphate of ammonia is a nitrogenous fertilizer which may upset the balance of growth if applied too heavily.

Sulphate of ammonia 2½oz/sq yd (70g/sq m)
Flowers of sulphur 2½oz/sq yd (70g/sq m)
Peat 3lb/sq yd (1.5kg/sq m)
Manure 6lb/sq yd (3kg/sq m)

it is no easy option at the construction stage.

There is even a case for carpeting the whole of a small garden with a simple ground cover, and making that the main element of design. During the preparation of this book the author walked along a road in a small town in Germany, admiring the different and sometimes ingenious ways the householders had made the most of their very small front gardens. Attractive though most of them were, just one stood out from a distance yet it was the simplest garden of them all. No hedge, no wall, no flowers, and no paving: just a solid carpet of *Pachysandra terminalis,* with a single small weeping tree set off-centre. It almost certainly needed the least maintenance of all the gardens in the road, it would obviously look smart the year round (even in winter the outline of the bare branches of the weeping tree would provide a focal point), and above all it was striking in its simplicity.

A garden like the one just described is not to everyone's taste (a plant enthusiast would find the lack of variety boring). But it demonstrates two points: ground cover can be used with great simplicity for a pretty effect without the effort that usually goes with gardening; and ground cover can be used with a sense of design that can have an impact equal to hard landscaping such as paving.

GETTING OFF TO A GOOD START

There is only one chance to get the ground preparation right. No amount of feeding and aftercare will improve the soil structure significantly once ground cover is established. Because ground cover plants are typically low maintenance (once established), and they are generally tough plants, it is often wrongly assumed that they will grow with the minimum of soil preparation. In fact they may need preparation as thorough as for, say, roses or dahlias. This is especially true for plants that are not in their natural habitat. Ground cover in nature flourishes with no adjustment to the soil, but it is provided by the plants most adapted to those conditions, and may not be particularly decorative for your specific use. It is especially important to prepare the ground properly if the planting is under trees, or other relatively dry or inhospitable places.

If you are planting a wide-spreading plant such as a prostrate juniper, it makes sense to improve the soil only in the localized planting area, to place the ground cover in the best soil and yet maintain the weeds at a disadvantage.

Usually, however, the whole area is treated, and there are few plants or soils that will not benefit from plenty of humus worked into the top 6-8in (15-20cm) of soil. It's usually best added after the weed control programme has been completed, and the best way to be sure of applying enough is to spread a 2-3in (5-8cm) layer of garden compost, well-rotted manure, spent hops, or old mushroom compost (or any other locally-available, bulky organic by-product), and fork this into the top 6in (15cm) or so. If you have to adjust the pH by adding lime, separate this job from the application of animal manures by a month or more (though if the soil is to be made more acid, manures and other organic material will help).

Organic material, which will eventually be broken down into humus, does not provide a large quantity of nutrients in the short-term. Unless planting one of the few plants that actually prefer an impoverished soil, rake in a balanced fertilizer at about 1-2oz per sq yd (30-65g per sq m). Growmore is popular in Britain, and this is a 7:7:7 formula of NPK (the bag will give this information – N stands for nitrogen, P for phosphorous, and K for potassium). A professional

A collection of beautifully textured ground cover plants. This is in Wuppertal botanic garden in West Germany, but a similar effect could be achieved in a modest-sized garden, instead of a lawn and flower beds.

Opposite: One of the autumn-flowering heathers, *Calluna vulgaris* 'Elsie Purnell'.

soil analysis will give more precise information on the nutrients most needed, but many fine gardens are achieved with the use of a balanced fertilizer alone.

Only if you garden on soils of an extreme kind (very acid or very alkaline for instance), are you likely to have major soil-induced deficiencies of particular nutrients. You can try to amend the soil to bring it to a more neutral level, but it's an uphill struggle that's not to be recommended if you are looking for a low maintenance (and low aggravation) garden.

WEED CONTROL

Weed-infested ground cover becomes an eyesore rather than an attraction, and once difficult perennial weeds are established among the plants, control is extremely difficult. A ground cover plant will win the battle with the weeds, once established. Even the best will require help initially.

Some gardeners have an aversion to chemicals in the garden, but herbicides make the job so much easier that it is worth considering their use initially even if you resort to hand weeding afterwards.

Land that is totally overgrown may need a total weedkiller, such as glyphosate, applying a few weeks before cultivation begins. It is difficult to dig ground where tall weeds are growing, and there's the chance of chopping up the roots of pernicious weeds such as couch grass *(Agropyron repens)*, also known as quack grass, and horsetails *(Equisetum spp)*, which in effect propagates them. You will then have to wait until they produce enough growth again for systemic insecticides such as glyphosate to work.

The advantage of glyphosate in comparison with older total weedkillers such as sodium chlorate is the ability to plant almost immediately, though it's best to wait for a week or two to ensure that the weedkiller has worked its way through the plant and killed the roots as well as the top growth.

Once deep-rooted perennial weeds have been killed it is comparatively easy to deal with annual weeds and the seedlings of perennials – many will germinate once the ground is cultivated, but nothing more dramatic than a hoe is necessary to control them. If you want to use chemical control for these, there are weedkillers that kill the top growth quickly and leave the soil free for almost immediate planting (in Britain, paraquat with diquat is used).

There are soil fumigants that will kill many of the dormant weed seeds as well as the growing weeds. Calcium cyanamide and metham are used in the States, but in Britain, soil fumigants like this are not used by amateurs.

If you don't want to use weedkillers, there is no option but to dig out as many roots of perennial weeds as possible, then keep hoeing the soil as new shoots or seedlings emerge. This may have to be done over a period of weeks or months, and it is better if the soil is not repeatedly dug, as more and more weed seeds will be brought to the surface. If the weather is dry, water the area to encourage seedlings to germinate quickly.

As the absence of light will eventually kill all weeds, a mulch of black plastic sheeting is another very effective non-chemical approach. If you can leave it in position for six months or even longer, the ground will be ready for cultivation, but of course this will bring more seeds to the surface to germinate. It's better to leave the black plastic in position, and plant through it. There's the advantage that you don't have to wait, provided you keep the planting holes small for the plants so that bare soil is not exposed.

The plants will in time cover the plastic, which will eventually disintegrate, but in the meantime you can improve the appearance by applying a covering of gravel or pulverized bark mulch.

PLANTING

Container-grown plants can be planted at any time provided the ground is not frozen or waterlogged, but spring and autumn are still sensible times. (If you are using weedkillers to clear the ground, some of these are best applied when the weeds are growing vigorously, often early summer, and this may affect your planting time.) Always water regularly if the ground is dry, but be especially prepared to do this if planting during a dry period.

Make the holes wide enough to take the plants without cramping them, and if planting a shrubby plant (or perhaps a prostrate conifer), tease out a few roots from the root-ball and spread these out in the planting hole, without disturbing the main mass of roots.

Some herbaceous plants may be sold bare-rooted, and it is important to plant these quickly before they dry out (keep them covered with damp newspaper while they are waiting to be planted – the sun or a strong wind will dry out fine roots quickly).

Try to plant bare-root plants at the same level that they were in the ground before (you should be able to see the soil mark), but the root-balls of container-grown plants are generally best just covered with the garden soil (it reduces the chance of the root-ball drying out).

Spacing depends on how quickly you want the plants to meet, as well as the actual size and ultimate spread of the plants. Optimum spacings for ground cover are given for each plant, but you can adjust these with moderation to suit budget and patience. If planting small plants closely, bear in mind that it's a good idea to leave enough space between each plant to be able to hoe easily.

It's useful to space the plants out on the ground before doing the actual planting, as it's easier then to make minor adjustments if the plants don't quite fill the area allocated. The cover will be better if you plant in staggered rows rather than rows that run parallel in both directions.

Planting a steep slope needs special care. Soil erosion, and the run-off of water, can leave the plants vulnerable and the slope unstable. Slight terracing (which will not be noticed once the plants have grown) will minimize some of the erosion problems and help the plants become established easily. But if the bank is really steep, try covering it with black plastic sheeting (jute netting is sometimes used in the USA), and plant into this. The bank will remain stable, with the minimum of soil erosion, until the plants have control.

PLANTING A SLOPE
A sheet of plastic helps to maintain the stability of a soil bank while ground covers become established. Make small holes in the plastic at the required planting distance and insert the young plants carefully.

It comes as a surprise to some gardeners to learn that conifers can be used for ground cover. This picture shows two suitable prostrate kinds – *Juniperus horizontalis* 'Douglasii' and (behind) *Juniperus communis* 'Effusa'.

Left: *Lonicera pileata*, a good low-maintenance shrubby ground cover for a border.

AFTERCARE

Watering is the most important after-planting care for all plants. Once established their roots will penetrate the soil and search for water, but in the early stages irrigation is needed to keep the plants growing rapidly and without check. It's worth using a sprinkler to ensure that they obtain enough moisture, as you would when establishing a new lawn.

Mulching helps by conserving soil moisture and by suppressing weeds, but organic mulches such as compost or peat need to be at least 2in (5cm) thick to suppress weed growth efficiently. A few weed seeds may fall on these mulches and germinate, but they are usually very easy to pull out.

Pre-emergent herbicides such as simazine are very useful for applying between the plants, as they will stop seedlings germinating for many months. But apply with care, and follow the manufacturer's instructions. It is probably best to avoid using them on a slope, where the chemical may be washed onto the ground cover plants or perhaps onto an adjoining lawn. It is also important that the ground cover plants' root-balls are well covered with soil, to reduce the risk of the weedkiller affecting the plants. This kind of weedkiller can be very useful, but must be applied carefully and according to instructions.

Feed the plants with a balanced fertilizer again in the first spring after planting, and if the plants look as though they are struggling, a further application later in summer may be necessary. In dry areas, such as beneath trees, be prepared to water regularly for the second season too.

If growth seems particularly slow, perhaps because the ground cover plants are sited where they are in competition for nutrients from neighbouring trees and shrubs, it is worth foliar feeding once or twice to give the plants a boost.

Regular weeding is, of course, essential until the ground covers have clothed the ground. Hoeing and hand weeding may not easily get rid of very difficult weeds such as bindweed. Be prepared to paint the leaves of these with a glyphosate gel, and repeat if regrowth occurs. This will eventually kill them.

Some woody plants may need pruning in the early years, to encourage a dense, bushy habit. Others, such as heathers, will need trimming with shears after flowering so that the plants remain compact and tidy. It may even be possible to mow a few ground covers with the blades set high (or clip them with hand shears) to encourage close, carpeting growth.

WHAT TO GROW WHERE

The following pages contain summaries of a selection of the most useful ground cover plants, suitable for use in Britain and other parts of the world with a similar climate. In the warmer parts of the United States, for instance, more colourful flowers like osteospermums and gazanias are used, but these are not an option for colder areas where they are treated as half-hardy plants. Annuals have not been included either, though for temporary ground cover, perhaps among perennials that are becoming established, they have their uses.

There are many other dependable ground cover plants not included in this list, so there is plenty of room for experimentation. But those described below are among the most generally reliable, and some of the best to start with.

Although listed alphabetically by Latin name, common names which are widely used have also been given. Where the common name generally used is different in the USA, this name has been given in brackets.

Acaena novae-zelandiae
NEW ZEALAND BURR

Soil	Site	Speed	Spacing
Well-drained	Sun	Fast	18-24in (45-60cm)

The New Zealand burrs (there are a few useful species, see below) are carpeting colonizers with small fern-like leaves, silvery green in summer turning bronze towards winter. Unfortunately they do not provide winter cover. The slender stems root as they spread, providing a quick dense cover that will suppress annual weeds and seedlings. But careful weeding is necessary to avoid perennial weeds becoming established during the first season or two, and grass seedlings may become established in spring before the new growth provides adequate cover.

The flowers, which generally appear between mid and late summer, are insignificant. But the russet-brown spiny burrs that follow are decorative.

A. novae-zelandiae, and the other species mentioned below, grow to about 2-3in (5-8cm) in height, and the effect is of a blue-green ground-hugging carpet (studded brownish-red when the burrs are present).

Uses Not a good choice when the prime object is weed suppression for a difficult spot, but good for an open position where you want a ground cover through which you can grow small bulbs such as crocuses and dwarf spring irises. Good for growing up to and among paving, and as ground cover at the front of a rock garden. It does well on poor and dry soils, and requires good drainage.

Other species *A. buchananii*, *A. caesiiglauca*, and *A. microphylla* are among the other species that are suitable (there are also hybrids). They vary mainly in the green or blue shade of the foliage; all are worth trying.

Acaena microphylla

Achillea tomentosa
WOOLLY YARROW

Soil	Site	Speed	Spacing
Well-drained	Sun	Fast	6-12in (15-30cm)

There are achilleas that are lawn weeds, some that are tall border plants, and others compact rock plants. The one to consider as a carpeting ground cover is *A. tomentosa*, an evergreen with feathery grey-green foliage. It is topped in mid and late summer with clusters of small yellow flowers; pretty in flower, but you need to trim the dead heads off afterwards. Although quite a bold plant, reaching 6-9in (15-23cm) high, it is generally best for quite small areas. If you want a dense carpet of foliage, mow it to about 2in (5cm), and sacrifice the flowers.

Uses It can be used as a lawn substitute provided you do not subject it to heavy wear. It also looks good as a ground cover at the foot of a rock garden, perhaps to link this with a lawn. Useful for poor, dry soils, provided the position is sunny.

Aegopodium podagraria 'Variegatum'
VARIEGATED GROUND ELDER
(SILVER-EDGED GOUT WEED)

Soil	Site	Speed	Spacing
Dry	Sun or shade	Fast	24in (60cm)

As ground elder is a pernicious weed of British gardens, which many gardeners spend hours trying to eradicate, this is included with some trepidation. But it's an easy-to-grow plant for difficult situations, and is planted as a ground cover in the USA. The variegated form is actually very attractive, grey-green edged white, and it has its uses in the wilder areas of the garden. It grows to about 12in (30cm) high.

The plant is herbaceous, and it will not provide winter cover, but the spring and summer growth is so vigorous that it will suppress weeds effectively.

Trim off the flowering stems as they appear. They will detract from the foliage effect, and if allowed to flower the seedlings can be invasive. Try mowing two or three times in a season with a rotary mower set as high as possible.

Uses Try this plant at the edge of a woodland area, or to carpet between shrubs in a large border. Not suitable for use where a compact, well-behaved plant is required, though it can be used successfully in a strip of ground between house and path, where it has firm boundaries to control its spread.

Ajuga reptans 'Rainbow'

Ajuga reptans
BUGLE (CARPET BUGLE)

Soil	Site	Speed	Spacing
Moist	Sun or partial shade	Medium	9-15 in (23-38cm)

One of the best ground covers for a moist soil: widely available, inexpensive, colourful. Although grown primarily for foliage effect (which is very colourful in some of the varieties), the blue flowers are also an attractive feature in spring. The plants spread by runners, to form a dense mat that is evergreen except in cold areas. Out of flower, the foliage forms a carpet 2-4in (5-10cm) high; as the flowers appear they bring the height of the plant to about 6in (15cm).

Choose one of the attractive hybrids or varieties. Among the best are: 'Atropurpurea' (reddish-purple; it may be sold as 'Purpurea'), 'Burgundy Glow' (shades of wine-red with touches of white and pink), 'Rainbow' (bronze, pink, and yellow variegation), and 'Variegata' (green edged creamy-yellow).

Ajugas do need occasional attention. The colonies tend to become bare in the centre if the soil is too dry or impoverished, but you can easily divide plants around the edge and replant the centre. Apply a general fertilizer once a year in spring or summer, watering it in thoroughly, to keep the growth lush and healthy.

To keep the plants looking neat, trim off the dead flowers with hand shears, or go over the plants with a lawnmower with the blades set high.

Uses An ideal plant to use in crevices among paving, or in bold drifts and ribbons to fill the gaps between a border and the lawn or a path (but be warned: it may encroach into the grass and become a weed).

Alchemilla mollis

LADY'S MANTLE

Soil	Site	Speed	Spacing
Ordinary	Sun or partial shade	Medium	15-18 in (38-45cm)

A clump-forming herbaceous plant that is equally effective as individual plants in the herbaceous border or as a summer ground cover. The large, lobed pale green leaves are attractive in their own right, and catch the morning dew drops which they hold like glistening jewels. The feathery sprays of tiny yellowish-green flowers are a charming bonus in early and mid summer. Height is 12-18in (30-45 cm).

Self-sown seedlings may become a problem if grown in borders.

Uses Because the plants die back to ground level for the winter, the lady's mantle is best regarded as a decorative ground cover for use in association with other plants. Try it as a summer carpeter for roses, or to edge a herbaceous or shrub border. It can look good in association with lavenders.

Anthemis nobilis

CHAMOMILE

Soil	Site	Speed	Spacing
Light	Sun or partial shade	Medium	9-15 in (23-38cm)

One of the best known alternatives to grass (see page 37), chamomile has been used for lawns in Europe for hundreds of years. This says much for its qualities as a ground cover, though its merits as a 'lawn' are more open to question.

Its fern-like foliage forms a mat about 3-10in (8-25cm) high if not clipped, and unless you choose the non-flowering form called 'Treneague', there will be masses of small, white, daisy-type flowers in summer. The leaves are aromatic if crushed, which is one reason for its popularity as a scented lawn. It should be mown to keep it compact and dense if you intend to walk on it (it will not tolerate heavy use).

Chamomile is not easy to establish, and is apt to die out in patches (these can be replanted with cuttings), and weeds can be a problem.

Uses Not a serious contender for a main lawn, but worth considering for a small ornamental lawn that is mainly for show and interest. It can make a useful ground cover perhaps with stepping stones to take most of the wear.

Armeria maritima
THRIFT, SEA PINK

Soil	Site	Speed	Spacing
Well-drained	Sun	Slow	9-12 in (23-30cm)

A distinctive evergreen that forms grassy tufts of foliage, studded with small pink, red, or white drumstick flower heads in late spring and early summer. In warm areas it will also flower intermittently throughout the summer. There are named forms such as 'Alba' (white), and 'Vindictive' (deep rose-red). Most have foliage to about 6in (15cm), with the flowers adding a little extra height.

In terms of flower display, this is one of the most attractive ground covers, and an established planting of thrift makes a textured, grass-like carpet.

Armeria maritima

It is a good choice for a coastal area, but does well in most areas and can thrive on relatively poor and dry soils. Go over the plant with shears after flowering, to remove the old flower heads.

Old clumps tend to die out in the centre, and may require dividing and replanting occasionally.

Uses Try it for an area that you can walk through on stepping stones. Or use it to provide ground cover where an attractive carpet is more important than weed control (it will take some years for the clumps to knit together to cover the ground completely). Useful as an edging.

Asarum
WILD GINGER

Soil	Site	Speed	Spacing
Moist	Shade	Medium	9-12 in (23-30cm)

Not a well-known plant, but very useful for a humus-rich moist soil in shade. Despite its common name, it is not really a ginger – this alludes to the pungent smell of the roots and foliage when bruised, which vaguely resembles ginger. The unusual bell-shaped flowers in spring are hidden by the leaves, and the plant is grown for its foliage effect. The shiny heart-shaped leaves are about 2-6in (5-15cm) across, and the plant spreads by creeping underground stems.

Slugs and snails can be a problem, and the plants will need regular summer watering if the ground is dry.

Uses Best in shade, in moist, organic, slightly acid soil, and for that reason often good beneath rhododendrons.

Other species There are other evergreen species to try, including *A. caudatum* and *A. shuttleworthii*, though the cover may not be so dense.

Aubrieta deltoidea

Astilbe chinensis 'Pumila'

SPIRAEA (FALSE ASTILBE)

Soil	Site	Speed	Spacing
Moist	Sun or shade	Medium	12-15 in (30-38cm)

This delicate-looking yet robust plant is smaller than the border and pondside astilbes, with pretty pink plumes in late summer over divided, feathery foliage that unfortunately only offers summer cover.

This astilbe grows little more than 6-12in (15-30cm) high, but spreads more readily than the hybrid types mentioned below.

Uses Best for small drifts rather than for covering a large area, especially at the front of a border. The large-flowered types mentioned below associate well with water and make excellent poolside ground cover (though with man-made ponds the ground beside them may be very dry, and astilbes like moist ground, so be prepared to water).

Other species There are many fine hybrids, which reach about 2-3ft (60-90cm) in flower, with plumes of white, pink, or red flowers in late spring and into summer. A handful of good ones are 'Deutschland' (white), 'Bressingham Beauty' (pink), and 'Fanal' (deep red).

Aubrieta deltoidea

Soil	Site	Speed	Spacing
Well-drained	Sun	Medium	10-12 in (25-30cm)

One of the most popular rock garden plants, aubrieta is seen in flower almost everywhere in spring. It is less often seen as a ground cover, but where you want to produce a sheet of vibrant colour there are few plants more capable. The drawback is that the plants become straggly after flowering and should be clipped over to remove the seed heads and encourage compact growth. Even so, summer cover can appear thin and sparse, and although the plants make a carpet of green 2-6in (5-15cm) high, there are much better choices for year-round interest. Grow just a small patch for early impact, but don't depend on it for interest the rest of the year.

Apply a balanced fertilizer once a year in spring or summer.

Uses Try aubrieta as a ground cover in front of a sunny stone wall, perhaps with plants inserted into crevices in the wall too. Can be effective around the base of a small deciduous tree provided there is not too much shade.

Bergenia

ELEPHANT EARS (LEATHER BERGENIA)

Soil	Site	Speed	Spacing
Any	Sun or shade	Medium	9-18 in (23-45cm)

One of the very best ground cover plants. Worth growing as individual specimens at the front of a border, and even more impressive massed where the big, leathery leaves provide complete and thorough ground cover. In spring, and possibly in summer, there's a bonus of striking sprays of pink (sometimes white or red) flowers, about 12in (30cm) tall.

The ground-hugging foliage sometimes turns purplish in winter.

The plants spread by thick rhizomes that run along the surface, yet they are never invasive. To keep them looking healthy and vigorous, however, it is worth dividing them every third year. Feed in spring.

Although they will grow in almost any soil, they prefer a moist, organic soil.

Uses These very adaptable plants make excellent ground cover in front of a border of shrubs, and can be very effective planted around the base of a specimen tree in a lawn. Try them as ground cover by a pond, where the large leaves will help to mask the edge of the pool.

Other species The two species most often grown are *B. cordifolia* and *B. crassifolia,* but there are other species, and many hybrids that are worth growing. It is best to avoid *B. ciliata,* and perhaps 'Ballawley', as the foliage may be damaged in a cold winter.

Bergenia

Calluna vulgaris
SCOTCH HEATHER

Soil	Site	Speed	Spacing
Acid, moist	Sun or partial shade	Medium	9-18 in (23-45cm)

This is one of the best-known heathers, and there are many different varieties ranging in size from about 4in (10cm) to 24in (60cm) and more, with evergreen foliage in colours that include gold, orange, bronze, grey-green, and green. The typical heather flowers come in shades of purple, pink, red, and white. Flowering time is from mid summer to late autumn.

Callunas do well in exposed and coastal areas, where they are especially useful ground covers. They do, however, demand an acid soil to do really well, though you can expect reasonable results on a neutral soil.

Clip over the plants with shears in spring, to maintain a compact, bushy habit.

Uses Heathers can be very useful for banks, but are also popular as ground cover in beds of dwarf conifers. Try them in bold drifts in the front of a sunny shrub border.

Campanula portenschlagiana
BELLFLOWER

Soil	Site	Speed	Spacing
Undemanding	Sun or partial shade	Medium	9-15 in (23-38cm)

The blue bells of the rock campanulas are well known, and there are many different species – some too delicate for ground cover, others almost too rampant. This species (which may also be listed as C. *muralis)* is one of the most suitable, growing to about 6in (15cm), with violet-blue bells over heart-shaped deep green foliage in mid to late summer. It will not give winter cover.

Uses A good plant to carpet the ground at the foot of a rock garden, or to grow along and between paving. Not suitable for ground cover for a large area.

Other species C. *poscharskyana,* the Serbian bellflower, has almost star-shaped blue flowers, on stems which tend towards trailing, in early summer. It grows as high as 24in (30cm) and is a more rampant plant, spreading by creeping runners and self-seeding freely.

Cerastium tomentosum
SNOW-IN-SUMMER

Soil	Site	Speed	Spacing
Undemanding	Sun	Fast	12-18 in (30-45cm)

If you are looking for a fast, easy, and inexpensive ground cover, this is one to try. The silvery-grey leaves are evergreen (though they sometimes look a bit tatty by the end of a hard winter), and in late spring and early summer the plants are simply smothered with white flowers. The creeping stems spread rapidly, and as the plants are easy to raise from seed they provide a cheap way of covering an area of ground quickly.

Snow-in-summer is not, however, an ideal long-term ground cover, as it tends to die out in patches with age. It forms a grey mat about 4-6in (10-15cm) high. Clip over the plants after flowering to keep them compact and to stimulate new growth.

Uses Ideal for quickly clothing any area in good light, but other than for a relatively small patch it can become too dominant and is best replaced by choicer and more permanent alternatives in time. Good for dry situations. Becomes straggly in shade.

Cerastium tomentosum

Ceratostigma plumbaginoides
LEADWORT, PLUMBAGO

Soil	Site	Speed	Spacing
Undemanding	Sun	Medium	12-24 in (30-60cm)

This colonizer, which spreads by suckers, never invades in an obtrusive way, and it is a delightful ground cover with blue phlox-shaped flowers in late summer and well into autumn. The thickets are usually about 12in (30cm) high, but can be taller, and the foliage assumes a reddish tint in the autumn. Although usually cut down by cold weather, it will grow again in spring except where the winters are very severe. The plants are best trimmed back after the first hard frost.

In areas prone to heavy winter frosts it may be necessary to apply a winter mulch of straw or bracken, but it's worth the effort for the pretty blue flowers late in the season.

Uses Try it in front of shrubs, where it can clothe the ground and mould itself between the shrubs.

Convallaria majalis
LILY-OF-THE-VALLEY

Soil	Site	Speed	Spacing
Rich, moist	Shade or partial shade	Slow	9in (23cm)

This is perhaps better known as a fragrance than as a garden plant, yet it is easy to grow and an effective ground cover. It's a good choice for cold winter areas. It is, however, poisonous and is perhaps best avoided if you have small children.

The nodding white bells on 8in (20cm) stems fill the air with a delicate fragrance in spring. There is also a pink-flowered variety, but if you want something

Convallaria majalis

special try the variegated form, which looks charming all summer. Unfortunately the foliage dies down for the winter.

Annual mulching in late winter or early spring will conserve moisture and help to suppress early weeds.

The plants spread slowly by underground stems, and eventually make a dense carpet.

Uses Try planting a bed of lily-of-the-valley around a tree, or use a drift at the front of a shrub border.

Cornus canadensis

CREEPING DOGWOOD

Soil	Site	Speed	Spacing
Acid, moist	Shade or partial shade	Medium	12-15in (30-38cm)

If you thought that all dogwoods were twiggy shrubs, you'll be surprised by this ground-hugging herbaceous carpeter. It is strictly for moist, acid soils, ideally rich woodland soils. Given the right conditions it will clothe the ground with pale green leaves, and in early

Cornus canadensis

summer produce its white-bracted flowers above the canopy of leaves about 6-9in (15-23cm) high. In ideal conditions the flowers are followed by scarlet berries.

The plant spreads extensively by its creeping rootstock.

Uses A carpeter for fairly open woodland, especially beneath deciduous trees, or around other acid-loving shrubs such as rhododendrons.

Cotoneaster dammeri

(BEARBERRY COTONEASTER)

Soil	Site	Speed	Spacing
Undemanding	Sun or partial shade	Medium	24-48in (60-120cm)

Several prostrate cotoneasters are useful ground covers, especially attractive in autumn and early winter when the stems are usually studded with red or orange berries. If you are confined to choosing just one, it makes sense to select an evergreen such as *C. dammeri.* This ground-hugger makes a carpet about 6in (15cm) high, the stems covered with small white flowers in early summer and later sealing-wax-red berries. The rooting branches can spread rapidly to 10ft (3m) or more, but cover may still be sparse initially and hand weeding will be necessary for the first year or two.

Uses Ideal for carpeting banks, or for clothing the ground beneath taller shrubs and trees.

Other species Other prostrate evergreens suitable for use as ground cover include *C. congestus, C. conspicuus* 'Decorus', and *C. microphyllus. C. horizontalis,* the fishbone cotoneaster (rock cotoneaster), is deciduous but such an outstanding plant, with berries often lasting well into winter, that it must be considered along with the evergreens.

Cotula squalida

(NEW ZEALAND BRASS BUTTONS)

Soil	Site	Speed	Spacing
Rich, moist	Sun or partial shade	Medium	4-6in (10-15cm)

An unattractive name for an evergreen carpeter with fern-like green foliage, growing to about 2in (5cm). The plant bears masses of small yellow flowers in mid summer.

The creeping stems root as they grow, forming a thick carpet, which can be used as a lawn substitute (with the proviso that it does not receive heavy wear).

Water in dry weather, and apply a fertilizer once a year.

Uses Although fairly invasive, it is a useful ground cover through which to grow dwarf bulbs. A good ground cover to use around and between paths.

Dryas octopetala

MOUNTAIN AVENS

Soil	Site	Speed	Spacing
Well drained	Sun	Medium	12-15in (30-38cm)

A carpeting evergreen with crinkled dark green leaves, felted beneath. It grows to about 2in (5cm) high, on creeping and branching stems that root freely, eventually causing a spreading mat of dense growth. The white flowers in early summer are about 1in (2.5cm) across, and followed by fluffy seed heads in early autumn.

Uses Ideal as ground cover at the foot of a rock garden, but also useful for edging a path, or as an edging to a border.

Epimedium perralderianum

Duchesnea indica

INDIAN STRAWBERRY, MOCK STRAWBERRY

Soil	Site	Speed	Spacing
Undemanding	Sun or partial shade	Fast	15in (38cm)

A rather coarse plant, with leaves like those of a strawberry. There are yellow flowers in late spring and early summer, and small red berries, but these are not especially decorative. The foliage is not completely evergreen, but is retained into winter.

A plant to choose for difficult spots where you need something tough.

Uses Suitable for covering a large area quickly, perhaps in partial shade, where decorative value is not so important.

Epimedium grandiflorum

BARRENWORT

Soil	Site	Speed	Spacing
Moist	Shade or partial shade	Medium	9-12in (23-30cm)

The barrenworts are useful almost evergreen ground cover plants. Subtle and interesting rather than bright and brash, but very effective. This species is often listed as *E. macranthum*, and there are other species just as worthy of a place in the garden (see below). All have heart-shaped leaves, usually pale green but sometimes flushed brown and often assuming bronze or red tints in autumn. The small flowers appear in late spring or early summer, but are generally hidden by the foliage. Trim off the old foliage with shears in very early spring, to tidy the plants and give the flowers a chance to show.

'Rose Queen', with pink flowers, is a variety of *E. grandiflorum* often grown, but other species and varieties have flowers in shades of yellow, white, and lavender. Most types grow to about 9in (23cm).

Uses Good for planting around a specimen tree, such as a crab apple or Japanese cherry, in a lawn. Effective between shrubs, and plants such as roses.

Other species *E. perralderianum* (yellow flowers) is exceptionally good. *E. × versicolor* (pink or yellow) and *E. × youngianum* (there are both pink and white forms) are others worth looking for, but these won't give such good winter cover.

Erica carnea, syn. *E. herbacea*

HEATH, HEATHER

Soil	Site	Speed	Spacing
Acid or neutral	Sun or partial shade	Medium	15-24in (38-60cm)

You are likely to find this winter-flowering heather sold under the name *E. carnea* or *E. herbacea,* and there are many varieties, in shades of pink, white, and red. Their peak flowering time is mid and late winter, when other flowers are scarce. These low-growing evergreens need no introduction: they are justifiably among the most widely planted shrubs.

Most heaths and heathers prefer an acid (or at least neutral) soil, but this species has the additional benefit of being lime-tolerant (but don't try growing it on shallow chalk soils).

Most varieties will make a carpet about 12in (30cm) high, but some types are more compact, a few a little taller. There are varieties with bronze or gold foliage, and these are worth planting in drifts among the green-leaved varieties to provide additional interest when the plants are not in flower, though they flower over an exceptionally long period of several months.

Trim ericas with shears after flowering.

Uses Good for clothing banks, or for planting as a carpet in beds containing dwarf conifers.

Other species There are several other species that make good ground cover: two popular ones are *E. × darleyensis* (also winter-flowering) and *E. vagans* (flowers mid summer to early autumn; not so lime tolerant).

Euonymus fortunei

(WINTER CREEPER)

Soil	Site	Speed	Spacing
Undemanding	Sun or shade	Medium	24in (60cm)

This is a versatile evergreen, happy to grow up a wall as a climber but superb as a ground cover too. It is good for erosion control and tough into the bargain.

Euonymus fortunei 'Silver Queen'

Choose the variegated forms, which add that extra touch of colour. Varieties include 'Emerald 'n' Gold' (green edged with gold), 'Emerald Gaiety' (creamy white and green, tinged pink in winter). 'Coloratus' or 'Colorata' (purplish in winter) is popular in the USA.

Most varieties grow to 12-24in (30-60cm) with an equal or somewhat broader spread.

Although not a serious problem in Britain, in the States euonymus scale can be a big drawback in some areas.

Uses Good for banks, and as a carpet for the front of a shrub border. Try the green and gold varieties as a ground cover between red-barked dogwoods, where they will form a striking contrast with the red twigs in winter.

Galeobdolon argentatum

YELLOW ARCHANGEL

Soil	Site	Speed	Spacing
Undemanding	Shade	Fast	24-36in (60-90cm)

It's necessary to give a list of other names under which you may find this plant: *Lamium galdobdolon, Galeobdolon luteum,* and *Lamiastrum galeobdolon* are sometimes used. The species itself is a coarse weed, but the variegated form ('Variegatum') is very useful for wild areas where a fairly rampant ground cover is suitable. It grows well under trees where many other ground cover plants struggle, and the silver marbled foliage is quite attractive. Yellow flowers appear in early and mid summer, on plants about 12in (30cm) tall.

Uses For ground cover under trees and shrubs, where its generous spreading is not likely to over-run less vigorous plants.

Gaultheria procumbens

PARTRIDGE BERRY, WINTERGREEN, CHECKERBERRY

Soil	Site	Speed	Spacing
Acid, peaty	Partial shade	Medium	12-15in (30-38cm)

A creeping evergreen with small leathery leaves, studded with red berries in autumn and into winter. Small, pinkish-white bell-like flowers are a bonus in mid and late summer, though they are often hidden by the leaves.

Uses A useful ground cover around acid-loving shrubs such as rhododendrons and azaleas.

Geranium endressii

CRANESBILL

Soil	Site	Speed	Spacing
Undemanding	Sun or partial shade	Medium	12-15in (30-38cm)

Although the cranesbills are herbaceous and die down in the winter, many provide dense summer cover that suppresses annual weed growth and seedlings well. And as they are desirable border plants anyway, they are sure to enhance the garden when in flower.

G. *endressii* is one of the best for ground cover. It grows to 12-18in (30-45cm), and the deeply lobed foliage is topped with pretty pale pink flowers from early to late summer. 'Wargrave Pink' is a particularly good variety with clear pink flowers.

Uses Best in bold drifts filling the gap between a shrub or herbaceous border and the lawn.

Other species Several other species are suitable, but one of the best is G. *macrorrhizum,* which has hairy, fragrant leaves on plants of about 12in (30cm), and

Geranium sylvaticum 'Mayflower'

flowers that are generally magenta or white. It provides extremely dense summer cover. Others to consider include *G.* × *magnificum*, *G. phaeum* (good for shade), *G. sanguineum*, and *G. sylvaticum*.

Hebe pinguifolia 'Pagei'

Soil	Site	Speed	Spacing
Undemanding	Sun or partial shade	Medium	15in (38cm)

Although several of the compact and dwarf hebes can be used as ground cover, this carpeter is perhaps the best. It has a good ground-hugging habit, and silvery foliage that is always striking. The small white flowers that top the carpet in late spring and early summer are a bonus. Height is seldom more than 9in (23cm).

You might find the plant sold simply as *H.* 'Pagei' or as *H. pageana*.

Uses Superb as a ground cover for the front of a shrub border, lying between the shrubs and a path or lawn. Contrasts well with pale green or gold-variegated foliage, such as *Elaeagnus pungens* 'Maculata'.

Hedera colchica

PERSIAN IVY

Soil	Site	Speed	Spacing
Undemanding	Sun or shade	Fast	3-4ft (90-120cm)

Although less often grown than the common English ivy *(H. helix)*, the Persian ivy is a more striking plant, especially in the form *H. c.* 'Dentata Variegata'. It has large heart-shaped leaves splashed green, cream, yellow, and grey – a real stunner when well grown.

Although it makes a magnificent foliage climber, it will also carpet a large area of ground with its big evergreen leaves.

Uses Only suitable for a large area, perhaps beneath trees.

Other species *H. canariensis*, the Canary Island ivy

(Algerian ivy), is similar, and the white marbled 'Variegata' (also sold as 'Gloire de Marengo') is the one to grow for impact. Unfortunately it is more easily damaged than the previous species in bad winters and is not suitable for cold areas.

Hedera helix

COMMON IVY (ENGLISH IVY)

Soil	Site	Speed	Spacing
Undemanding	Sun or shade	Fast	24-48in (60-120cm)

The common ivy needs no introduction. But there are many varieties from which to choose, and some are better than others for ground cover. One of the best for this job is *H. helix hibernica*, the Irish ivy, which has dense, vigorous growth that provides quick cover (it is sometimes listed as a separate species – *H. hibernica*).

The variegated small-leaved ivies are less effective as ground covers for a large area, but don't be afraid to try any of them for a small piece of ground, perhaps around the base of a specimen tree in the lawn. You can let the ivy climb up the tree too if you wish – a bright one like 'Goldheart' (also known as 'Jubilee') can look stunning.

Uses Most useful for carpeting the ground beneath trees.

Helianthemum nummularium

SUN ROSE, ROCK ROSE

Soil	Site	Speed	Spacing
Well-drained	Sun	Medium	12-18in (30-45cm)

The hybrids of this charming plant come in shades of pink, red, and yellow, as well as white. They make spreading evergreen or semi-evergreen shrubs that seldom grow to more than 12in (30cm), and many have attractive greyish foliage that makes them interesting over a long period. It is the profusion of pretty flowers covering the plants in early and mid summer that is the real attraction, however, and if you clip the dead flowers off with shears you will probably get a second flush later in the year. This also keeps the plants compact and bushy. Even so they eventually become straggly and may need replanting after a few years.

There are many varieties, including doubles. The vigorous single varieties are best for ground cover. 'Wisley Pink' is particularly good.

Uses Ideal for clothing a sunny bank, especially on chalky (alkaline) soils.

Hosta

PLANTAIN LILY, FUNKIA

Soil	Site	Speed	Spacing
Undemanding	Sun or shade	Medium	15-18in (38-45cm)

In recent years hostas have become so popular that there should be no problem in obtaining a wide selection of species and varieties. Most of them will make good ground cover, albeit that they die down for the winter. The mass of usually large leaves ensures that those few weed seedlings that do germinate in spring are soon smothered.

There is such a bewildering array of hostas, many of them beautifully variegated, the choice must be a personal one. But if you want a starting point for ground cover, and don't mind doing a bit of searching for them, consider *H. ventricosa* 'Aureo-marginata', or the hybrids 'Ground Master', 'Hadspen Blue', or 'Shade Fanfare'. But any of the varieties more widely

Hosta plantaginea

available from garden centres can be used.

A couple of warnings are necessary: you must be prepared to control slugs and snails, which can ruin the display, and the ground must be weed-free to start with.

Uses Can be used in beds devoted to hostas, but these will look uninteresting for six months of the year. Best in drifts in front of shrubs, or in open woodland.

Hypericum calycinum

ROSE OF SHARON, AARON'S BEARD
(CREEPING ST JOHN'S WORT)

Soil	Site	Speed	Spacing
Undemanding	Sun or partial shade	Fast	18in (45cm)

This is one of those really tough ground cover plants that don't give weeds a chance. It's often maligned as being coarse and invasive and therefore not worth growing, yet in the right place it's ideal.

The plants grow to about 12in (30cm), spreading freely by underground shoots. The foliage is evergreen except in very severe winters, when it becomes semi-evergreen, and the ground always seems clothed. In mid and late summer it can be really spectacular, with its bright yellow flowers, more than 2in (5cm) across, each with a bold boss of stamens.

The effect is often better if the plants are sheared to the ground each spring, before growth starts, to stimulate fresh foliage.

Uses A good plant for controlling soil erosion on a bank. Useful for any area where its spread can be confined, perhaps around a tree set in an area surrounded by paving.

Hypericum calycinum

Iberis sempervirens
PERENNIAL CANDYTUFT

Soil	Site	Speed	Spacing
Undemanding	Sun or partial shade	Medium	12-15 in (30-38cm)

A flowering evergreen that is often seen in rock gardens, but less frequently used as a ground cover. It makes compact mounds about 9in (23cm) tall (more compact in the variety 'Snowflake' and 'Little Gem'), becoming a sheet of white flower heads in mid and late spring.

Trim back after flowering.

Uses A good edging for a path, or anywhere for small-scale flowering ground cover. Associates well with rock garden plants.

Iberis sempervirens

Juniperus
JUNIPER

Soil	Site	Speed	Spacing
Well-drained	Sun	Medium to slow	30in (75cm)

Not everyone thinks of conifers when it comes to choosing a ground cover, yet some of the prostrate types are among the very best where space is not a limitation. The junipers provide most of the best kinds. All those mentioned provide dense evergreen cover and require very little maintenance (they may benefit from some annual pruning to improve density, and in later years to avoid encroachment).

J. communis 'Repanda' (dark green) is dependable and grows to little more than 10in (25cm) high. It is among the many forms of *J. horizontalis*, however, that some of the finest ground cover conifers are to be found. Although there are many other suitable varieties, look for 'Blue Chip', also known as 'Blue Moon', (silver-blue summer foliage, blue-grey in winter), 'Bar Harbor' (grey-green, silvery in winter, flat-branching, spreading growth), and 'Glauca' (blue-green). *J. procumbens* 'Nana' (blue-green foliage) grows to about 12in (30cm) but can be rather slow.

Some of the taller but horizontally spreading junipers also make good ground cover where there is space and you need more height. *J. sabina tamariscifolia* makes a low-growing flat-topped bush to about 30in (75cm), and in the USA there is a variety of this ('Tamariscifolia No Blight') that is resistant to tip blight, a problem in some parts of America. For a really bold plant, try *J.* × *media* 'Pfitzerana Aurea' or the similar but more compact 'Old Gold', both of which have gold-tipped young foliage (lasting longer in the latter variety).

Uses All these junipers are best planted in a large area (otherwise they are best grown as single specimens),

Lamium maculatum 'Beacon Silver'

perhaps in a long border alongside a path, or as ground cover in a broad bed in front of taller conifers. Try them in island beds, with some taller contrasting conifers to provide height.

Lamium maculatum
DEAD NETTLE

Soil	Site	Speed	Spacing
Undemanding	Sun or shade	Fast	12-15in (30-38cm)

A very easy and attractive herbaceous ground cover with white-splashed leaves and white or purple flowers in spring and early summer. It makes a loose carpet about 9in (23cm) high, rooting as it spreads. 'Beacon Silver' has silvery-white foliage that makes a bright splash for a dull spot. 'White Nancy' has white flowers over a carpet of silvery-grey foliage, and is not invasive.

Trim the plants lightly once they have flowered, to keep them dense and bushy.

Uses Good where a quick and easy ground cover is required. The white forms are useful for brightening shaded corners.

Liriope muscari
(CREEPING LILY-TUFT)

Soil	Site	Speed	Spacing
Undemanding	Sun or shade	Slow	9-12in (23-30cm)

Useful for its late flowering season. The spikes of blue flowers, resembling large grape hyacinths, appear from the tufts of evergreen grassy leaves between late summer and mid autumn. The spreading clumps reach about 24in (60cm) in height, so this is not a compact ground cover, but the plants can compete with tree roots and other plants without being invasive.

Good for dry positions, and will grow in sun or shade, though in shade the growth is thinner.

Uses Good in front of shrubs or as an edging to a shady border.

Other species *L. spicata* is a better choice for very cold areas. It has lavender flowers and grows only to about 12in (30cm).

Liriope muscari

Lonicera pileata

Soil	Site	Speed	Spacing
Undemanding	Sun or shade	Medium	24-30in (60-75cm)

A foliage, shrubby honeysuckle, reaching about 24in (60cm) in height. It is a semi-evergreen shrub with a rather horizontal growth habit that makes it interesting if not spectacular. Translucent violet berries in autumn are an additional feature.

Uses Only suitable for use in a large garden, where a border or bed of ground cover is required, perhaps in front of a hedge or trees.

Other species The climbing honeysuckles, such as *L. periclymenum*, grown for their fragrant flowers, can be used as ground cover. They will scramble along the ground if there is no support, but they don't look tidy, and will require regular trimming.

Luzula maxima, syn. L. sylvatica

WOODRUSH

Soil	Site	Speed	Spacing
Undemanding	Shade	Medium	12-15in (30-38cm)

This is a ground cover plant perhaps more used in Europe than the States, and even then not common. But it has been included because it is so useful for difficult spots such as dry, shady areas. 'Variegata' is a form with white-edged leaves, otherwise the plants resemble mats of coarse grass.

Uses Dry, shady areas beneath trees.

Luzula maxima

Lysimachia nummularia
CREEPING JENNY, MONEYWORT

Soil	Site	Speed	Spacing
Undemanding	Sun or shade	Fast	12-15in (30-38cm)

Although usually described as evergreen, this plant will lose its leaves in winter in all but the mildest areas. But it's a wonderfully adaptable plant that is ideal for cascading down walls and spreading its way between all kinds of plants as a ground cover, hugging the ground as it goes.

Although it thrives in moist soils, and does well in boggy ground at the edge of a pond or stream, it will also perform well in dry, sandy soils. The small, rounded leaves provide good summer cover, though they are no match for difficult weeds. The main attraction is the mass of bright yellow flowers appearing in early and mid summer. 'Aurea' has golden foliage, but it is less vigorous and the leaves may be scorched in full sun.

Uses Ideal as a quick carpeter, where a specially attractive display is more important than good weed suppression.

Pachysandra terminalis
(JAPANESE SPURGE)

Soil	Site	Speed	Spacing
Neutral/acid	Shade or partial shade	Medium	9-12in (23-30cm)

This comes close to the ideal ground cover for shady areas, perhaps beneath trees. It makes a level evergreen carpet of foliage about 8in (20cm) high, covering the ground fairly quickly without becoming invasive, and creating an area of very uniform 'texture'. The variety 'Variegata', with white variegation, is more attractive, and marginally less vigorous. One of the most widely planted ground covers, with good reason.

They do not do well on alkaline soils.

Uses Ideal for difficult spots, such as the wooded edges of a large garden, but don't be afraid to use it in more open positions.

Polygonum affine

KNOTWEED (HIMALAYAN FLEECEFLOWER)

Soil	Site	Speed	Spacing
Well-drained	Sun	Medium	9-12in (23-30cm)

A really decorative plant, worth growing at the front of the border, or on low walls or in gravelled areas, even if not used for ground cover. In fact as ground cover it can tend to die out in patches, but it's well worth growing in an area where you don't mind doing some occasional replanting.

It forms a creeping mat of bright green foliage to about 6in (15cm), the spreading shoots rooting as they grow. The leaves usually become bronze-crimson in autumn, and are lost for the winter.

The small pink pokers bring the flowering height to about 10in (25cm). 'Donald Lowndes' is a good deep pink form, 'Darjeeling Red' a nice crimson. The flowers appear over a long period from mid summer to early autumn.

Uses Attractive around the base of small ornamental specimen trees in a lawn, or at the front of a sunny shrub border or rock garden.

Other species *P. vacciniifolium* is sometimes slow to establish, but worth waiting for as the small pink flower spikes are very attractive in early and mid autumn. It grows to about 4in (10cm). Some of the larger polygonums can be rather coarse.

Polygonum affine

Potentilla fruticosa

SHRUBBY POTENTILLA, SHRUBBY CINQUEFOIL

Soil	Site	Speed	Spacing
Fertile	Sun	Medium	24-30in (60-75cm)

These compact shrubs are certainly not carpeters, but where something growing to about 24in (60cm) is acceptable they have the merit of attractive flowers over a long period. Although deciduous, they knot together to form a dense summer cover.

There are several good named varieties suitable for ground cover. Those to look for include 'Longacre' (bright yellow), 'Elizabeth' (canary yellow) and 'Mandschurica' (silvery foliage, white flowers). The first two may be found listed simply as hybrids, and the latter may be catalogued as *P. dahurica* 'Manchu'. All can flower on and off all summer.

Uses Good for the front of a border, or for clothing a sunny bank.

Other species The shrubby potentillas should not be confused with the herbaceous types, some of which are also used as ground cover.

Prunus laurocerasus 'Otto Luyken'

Prunus laurocerasus
CHERRY LAUREL

Soil	Site	Speed	Spacing
Well-drained	Sun or shade	Medium	36-48in (90-120cm)

Laurels may appear to be boring evergreens, but there are a couple of varieties of cherry laurel that make very practical and attractive ground cover. 'Otto Luyken' is a compact shrub with shining green leaves, and attractive candles of white flowers in mid spring; 'Zabeliana' has a low, horizontally branching habit. Neither normally reach much more than about 48in (120cm) in height.

Uses Most suitable for a large garden, perhaps to soften the line between a row of trees and a path, or in front of a shrub border. Will do well beneath trees.

Pulmonaria
LUNGWORT

Soil	Site	Speed	Spacing
Moist	Shade or partial shade	Medium	12in (30cm)

Three pulmonarias are widely used for ground cover: *P. angustifolia, P. officinalis,* and *P. saccharata.* The first of these has plain green leaves, the other two have silver-spotted or blotched foliage that makes them particularly attractive as ground covers. The flowers are usually in shades of blue, though they may open pink, appearing about mid spring. There are varieties with pink or white flowers.

The foliage will generally disappear in winter, but the leafy rosettes provide good summer cover.

Uses Good ground cover for a shady border, or in moist ground beneath trees.

Rosa
ROSE

Soil	Site	Speed	Spacing
Undemanding	Sun	Medium	24-48in (60-120cm)

Nearly everyone loves a rose, and the idea of a carpet of roses has an obvious appeal. Provided you do not expect too much in the way of ground cover and weed suppression, they can be successful.

Rosa 'Grouse'

There are two main approaches to ground cover with roses: you can use a prostrate, creeping type and keep it low by the removal of older growth; or you can use arching shrubs that may reach 36in (90cm) or more. In both cases routine pruning should include removing old woody growth, and shortening any branches that tend to grow vertically.

Specialist catalogues list varieties suitable for ground cover, but three worth looking for are 'Max Graf' (pink flowers with gold centres; there is also a 'Red Max Graf'), 'Nozomi' (blush-white), and *R. wichuraiana* (very fragrant, small white flowers; almost evergreen in mild areas).

There are many others suggested as ground cover, but to avoid false expectations, it's worth trying to see them growing before you buy – out of flower as well as in flower. In all cases you will have to be prepared to weed for at least two seasons.

Uses Perhaps best for clothing banks.

Sarcococca humilis
CHRISTMAS BOX

Soil	Site	Speed	Spacing
Moist	Shade or partial shade	Medium	12-15in (30-38cm)

Not only a good ground cover, but a plant with winter fragrance too. It is a small leafy shrub with upright habit, growing to about 24in (60cm), and spreading by underground runners. The flowers come in mid or late winter, and although rather inconspicuous have the merit of being fragrant.

The plants respond to annual feeding.

Uses Useful for mass planting in a shady position, or in front of shrubs such as rhododendrons and camellias.

Stachys byzantina 'Silver Carpet'

Stachys byzantina, syn. *S. lanata*
LAMB'S TONGUE, LAMB'S EARS

Soil	Site	Speed	Spacing
Well-drained	Sun	Medium	12-15in (30-38cm)

An evergreen (actually grey) carpeter with narrow woolly leaves that give the plant its common names. By the end of a hard winter, the foliage will look much the worse for wear, but it soon recovers in spring. 'Silver Carpet' is a good variety to choose if you want just foliage effect, but if you find the whorls of small purple flowers in summer attractive, choose one of the other forms. Most of the summer it makes a carpet of silvery-grey about 6in (15cm) high, but the flower spikes take it up to 18in (45cm).

Trim the plants with shears in early spring if the foliage is in poor condition, and feed.

You may also find this plant listed as *S. olympica*.

Uses Its colour makes it an excellent choice to plant where there are mainly green background shrubs. But it also associates well with silver-leaved herbaceous plants such as some of the dianthus, and the blue-flower catmints (nepetas). Good for a hot, sunny spot.

Symphytum grandiflorum
COMFREY

Soil	Site	Speed	Spacing
Undemanding	Shade or partial shade	Fast	12in (30cm)

A rather leafy, coarse plant, but effective in the wild garden, or beneath trees, covering the ground rapidly. The yellow, pink, or blue flowers appear in spring, and although they add interest are not especially attractive. Being herbaceous it dies down in winter, but the summer growth is so dense that it is still an effective weed control. It generally makes a carpet about 12in (30cm) high, but may be taller in shade.

Uses Good for woodland or beneath trees in the wilder parts of the garden.

Tiarella cordifolia

Thymus serpyllum
THYME, CREEPING THYME

Soil	Site	Speed	Spacing
Well-drained	Sun	Medium	9in (23cm)

The thyme used for ground cover is the creeping species *T. serpyllum,* not the more bushy *T. vulgaris,* which is the culinary herb. It forms a close, ground-hugging carpet. The tiny leaves are aromatic, which makes thyme a popular choice to grow between paths and stepping stones. In summer there are clusters of tiny purple, white, pink, red, or lavender flowers, much appreciated by bees.
Uses A good ground cover among stepping stones. It can be used as a lawn substitute.
Other species The woolly thyme, *T. s.* 'Lanuginosus' (also found as *S. praecox pseudolanuginosus, S. pseudolanuginosus,* and *T. lanuginosus)* is worth considering as a ground cover, despite the confusion over nomenclature! It has hairy grey-green foliage and spreads rapidly by long trailing stems, rooting as they grow and hugging the ground.

Tiarella cordifolia
FOAM FLOWER

Soil	Site	Speed	Spacing
Moist	Shade	Medium	9in (23cm)

The lobed, hairy leaves of this useful ground cover make a carpet about 4in (10cm) high, topped in late spring and early summer with fluffy spikes of white flowers. The foliage turns bronze in autumn, and new leaves appear in spring.
Uses Excellent for a shady position in a humus-rich woodland soil. Also good for a semi-woodland area.

Vinca minor 'Aureomarginata'

Vinca minor

LESSER PERIWINKLE

Soil	Site	Speed	Spacing
Undemanding	Shade or partial shade	Fast	18in (45cm)

One of the most popular and reliable ground covers. The evergreen foliage, on rather sprawling stems that root as they trail, is never unattractive, but it's the typical blue periwinkle flowers that make it popular. There are also white and purple varieties, and variegated forms. All grow to about 6in (15cm).
Uses Ideal for shady banks, and for growing in the dappled shade of trees and shrubs.
Other species The other popular species is *V. major*, the greater periwinkle. It is a coarser plant, growing to about 9in (23cm), but otherwise similar. Again there is a variegated form.

Waldsteinia ternata

Soil	Site	Speed	Spacing
Moist	Sun or shade	Fast	12in (30cm)

An evergreen perennial that spreads rapidly by creeping, rooting stems, to form a carpet about 4in (10cm) high. It's a useful ground cover where something quick and easy is required. The yellow flowers in spring and early summer, often most profuse at the edge of the mat, are a bright and cheery bonus.
Uses A good ground cover among low-growing shrubs, or as a border to woodland paths, and for filling in those narrow strips between hedge and path.
Other species *W. fragarioides*, the barren strawberry, is similar, and often used instead of *W. ternata* in the USA.

GROUND COVER PLANTS FOR SPECIAL USES

Twelve to Grow Beneath Trees and Shrubs

Aegopodium podagraria 'Variegatum'

Convallaria majalis

Cornus canadensis

Duchesnea

Galeobdolon argentatum

Hedera

Hypericum calycinum

Liriope

Pachysandra

Prunus laurocerasus

Sarcococca

Vinca

Ten for Showy Flowers

Armeria

Astilbe

Aubrieta

Bergenia

Calluna

Campanula

Convallaria

Erica

Helianthemum

Hypericum calycinum

Ten for Dry Soil

Acaena

Achillea

Armeria

Campanula poscharskyana

Cotula

Hebe pinguifolia 'Pagei'

Helianthemum nummularium

Prunus laurocerasus

Juniperus

Thymus serphyllum

Twelve of the Best All-Round Ground Cover Plants

Ajuga

Alchemilla

Bergenia

Calluna

Cotoneaster dammeri

Erica

Euonymus fortunei

Hebe pinguifolia 'Pagei'

Hedera

Hypericum calycinum

Pachysandra terminalis 'Variegata'

Stachys byzantina

Ten for Shade

Aegopodium podagraria 'Variegatum'

Asarum

Epimedium grandiflorum

Hedera

Hosta

Galeobdolon argentatum

Moss

Pachysandra terminalis

Symphytum grandiflorum

Vinca

Five to Grow Between Stepping Stones

Anthemis nobilis

Armeria

Aubrieta

Cotula

Thymus serphyllum

Six for Erosion Control

Cotoneaster

Euonymus

Hedera

Hypericum calycinum

Juniperus

Vinca

INDEX

Page numbers in *italics* refer to illustrations.